Solutions Manual

for

Partial Differential Equations: An Introduction
Second Edition

THE WILEY BICENTENNIAL—KNOWLEDGE FOR GENERATIONS

*E*ach generation has its unique needs and aspirations. When Charles Wiley first opened his small printing shop in lower Manhattan in 1807, it was a generation of boundless potential searching for an identity. And we were there, helping to define a new American literary tradition. Over half a century later, in the midst of the Second Industrial Revolution, it was a generation focused on building the future. Once again, we were there, supplying the critical scientific, technical, and engineering knowledge that helped frame the world. Throughout the 20th Century, and into the new millennium, nations began to reach out beyond their own borders and a new international community was born. Wiley was there, expanding its operations around the world to enable a global exchange of ideas, opinions, and know-how.

For 200 years, Wiley has been an integral part of each generation's journey, enabling the flow of information and understanding necessary to meet their needs and fulfill their aspirations. Today, bold new technologies are changing the way we live and learn. Wiley will be there, providing you the must-have knowledge you need to imagine new worlds, new possibilities, and new opportunities.

Generations come and go, but you can always count on Wiley to provide you the knowledge you need, when and where you need it!

WILLIAM J. PESCE
PRESIDENT AND CHIEF EXECUTIVE OFFICER

PETER BOOTH WILEY
CHAIRMAN OF THE BOARD

Solutions Manual

for

Partial Differential Equations: An Introduction
Second Edition

Julie L. Levandosky

Steven P. Levandosky

Walter A. Strauss

John Wiley & Sons, Inc.

To order books or for customer service please, call 1-800-CALL WILEY (225-5945).

ISBN-13   978-0-470-26071-5

Printed in the United States of America

SKY10064267_010824

Printed and bound by Bind Rite Robbinsville, Inc.

Preface

This is the solutions manual for the Second Edition of the text, Partial Differential Equations: An Introduction" by Walter A. Strauss. We give detailed answers to about half the exercises. The manual is intended to be used by the student as a study guide in conjunction with the text itself. We hope that it will make the text more user-friendly.

The exercises have generally been chosen to be a representative sample. However, they purposely do include many of the more difficult and lengthy ones. In a few cases we have presented more than one method of solution. We have tried to be consistent with the notation of the text, including the numbering of the equations.

A few exercises have changed their numbering from the First Edition. These re-numberings are listed at the end of the manual.

We would appreciate readers pointing out any errors to us.

Table of Contents

Chapter 1

Section 1.1

1.1.2.

(a) $\mathscr{L}$ is linear.

(b) $\mathscr{L}$ is not linear, because of the uu_y term.

1.1.3.

(a) The equation is second-order, linear and inhomogeneous. It takes the form $\mathscr{L}(u) = -1$, where $\mathscr{L}(u) = u_t - u_{xx}$ is linear.

(c) The equation is third-order, nonlinear. The term uu_x is nonlinear.

1.1.5.

(a) The set is a vector space. Given any two vectors $[a_1, 0, c_1]$ and $[a_2, 0, c_2]$ in the set, their sum

$$[a_1, 0, c_1] + [a_2, 0, c_2] = [a_1 + a_2, 0, c_1 + c_2]$$

is also in the set, and the scalar multiple

$$c[a_1, 0, c_1] = [ca_1, 0, cc_1]$$

is in the set. Thus the set is a vector space.

(c) The set is not a vector space. Since both $[0, 1, 1]$ and $[1, 0, 3]$ are in the set, but their sum $[1, 1, 4]$ is not, since the product of the first two components is 1, not 0. So the set is not closed under addition. (It is closed under scalar multiplication, but that is irrelevant.)

1.1.7. The functions $1 + x, 1 - x$ and $1 + x + x^2$ are linearly independent. Suppose

$$a(1 + x) + b(1 - x) + c(1 + x + x^2) = 0$$

Then $cx^2 + (a - b + c)x + (a + b + c) = 0$. Since this must hold for every x, all three coefficients c, $a - b + c$ and $a + b + c$ must be zero. Since $c = 0$, the other equations imply $a = b$ and $a = -b$, which implies $a = b = 0$. So the only linear combination of $1 + x$, $1 - x$ and $1 + x + x^2$ which equals zero is the trivial combination.

1.1.10. Let u and v be any two solutions of the differential equation. Then

$$(u + v)''' - 3(u + v)'' + 4(u + v) = u''' - 3u'' + 4u + v''' - 3v'' + 4v = 0 + 0 = 0$$

and
$$(ku)''' - 3(ku)'' + 4(ku) = k(u''' - 3u'' + 4u) = k(0) = 0$$
so both $u + v$ and ku are solutions of the differential equation. Thus the set of solutions is closed under addition and scalar multiplication. The characteristic equation of the differential equation is $r^3 - 3r^2 + 4 = 0$, whose roots are $r = -1$ and $r = 2$ with multiplicity 2. Hence the solution set is spanned by $\{e^{-t}, e^{2t}, te^{2t}\}$. Since this spanning set is linearly independent, it is a basis for the set of solutions.

1.1.11. Since $u_x = f'(x)g(y)$, $u_y = f(x)g'(y)$ and $u_{xy} = f'(x)g'(y)$,

$$uu_{xy} = f(x)g(y)f'(x)g'(y) = f'(x)g(y)f(x)g'(y) = u_x u_y.$$

1.1.12. Since

$$\frac{\partial u_n}{\partial x} = n\cos nx \sinh ny, \qquad \frac{\partial^2 u_n}{\partial x^2} = -n^2 \sin nx \sinh ny$$

and

$$\frac{\partial u_n}{\partial y} = n\sin nx \cosh ny, \qquad \frac{\partial^2 u_n}{\partial y^2} = n^2 \sin nx \sinh ny$$

we have

$$\frac{\partial^2 u_n}{\partial x^2} + \frac{\partial^2 u_n}{\partial y^2} = -n^2 \sin nx \sinh ny + n^2 \sin nx \sinh ny = 0.$$

Section 1.2

1.2.1. The general solution is $u(x,t) = f(2x - 3t)$. The initial condition therefore implies $f(2x) = \sin x$, so $f(s) = \sin(s/2)$. Hence $u(x,t) = \sin(x - \frac{3}{2}t)$.

1.2.3. The equation is $\nabla u \cdot (1 + x^2, 1) = 0$, so the characteristics are solutions of

$$\frac{dy}{dx} = \frac{1}{1 + x^2}.$$

Integration yields $y = \arctan x + C$. Thus $u(x,y) = f(y - \arctan x)$, where f is any differentiable function. Some of the characteristics are shown in Figure 1.

1.2.7.

(a) The equation is $\nabla u \cdot (y, x) = 0$, so the characteristics must solve

$$\frac{dy}{dx} = \frac{x}{y}.$$

Separating variables yields $y^2 = x^2 + C$, so the general solution is $u(x,y) = f(x^2 - y^2)$. The initial condition implies $f(-y^2) = e^{-y^2}$, so $f(s) = e^s$ for $s \leq 0$. Thus one solution is $u(x,y) = e^{x^2 - y^2}$.

2

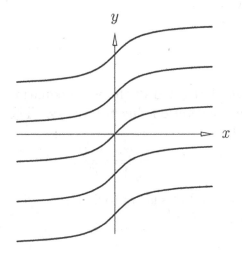

Figure 1: Characteristics of $(1+x^2)u_x + u_y = 0$.

(b) The auxiliary condition in part (a) determines $f(s)$ for $s \le 0$, and therefore the solution is determined uniquely only in the region $\{(x,y) \mid x^2 \le y^2\}$.

1.2.8. Making the change of variable $x' = ax + by$, $y' = bx - ay$ gives

$$u_x = au_{x'} + bu_{y'}, \quad u_y = bu_{x'} - au_{y'}$$

so the equation becomes

$$(a^2 + b^2)u_{x'} + cu = 0.$$

The solution of this ODE in x' is

$$u(x', y') = f(y')e^{-cx'/(a^2+b^2)},$$

where f is any differentiable function. In the original variables x and y,

$$u(x, y) = f(bx - ay)e^{-c(ax+by)/(a^2+b^2)}.$$

1.2.13. Let $x' = x + 2y$, $y' = 2x - y$. Then the equation becomes $5u_{x'} + y'u = x'y'$. This is a first-order linear ODE in x'. Using the integrating factor $\mu = e^{x'y'/5}$ it becomes

$$\frac{d}{dx'}\left(e^{x'y'/5}u\right) = \frac{1}{5}e^{x'y'/5}x'y'$$

so

$$e^{x'y'/5}u = \int \frac{1}{5}e^{x'y'/5}x'y' \, dx' = x'e^{x'y'/5} - \frac{5}{y'}e^{x'y'/5} + g(y'),$$

where g is an arbitrary differentiable function. This implies $u = x' - \frac{5}{y'} + e^{-x'y'/5}g(y')$. In the original variables (x, y) this becomes

$$u(x, y) = x + 2y - \frac{5}{2x - y} + e^{-(2x^2+3xy-2y^2)/5}g(2x - y).$$

3

Section 1.3

1.3.2. Proceeding as in the derivation of the wave equation, let ρ be the mass density of the chain and let g be the acceleration due to gravity. The horizontal component of the gravitational force is given by

$$\frac{u_x \rho g (l-x)}{\sqrt{1+u_x^2}}.$$

Assuming the oscillations are small this is approximately $\rho g (l-x) u_x$. Thus the net horizontal force on an interval $[x_0, x_1]$ is

$$\rho g (l-x) u_x \bigg|_{x_0}^{x_1} = \rho g \left[(l-x_1) u_x(x_1, t) - (l-x_0) u_x(x_0, t) \right].$$

By Newton's Law this equals

$$\int_{x_0}^{x_1} \rho u_{tt}(x, t)\, dx.$$

Differentiating with respect to x_1 gives

$$\rho u_{tt}(x_1, t) = \rho g \left[(l-x_1) u_x(x_1, t) \right]_{x_1}.$$

Thus

$$u_{tt} = g((l-x)u_x)_x.$$

1.3.4 Since the concentration is uniform in x and y, it is described by a function $u(z, t)$. Consider fluid within the region $z_0 < z < z_1$. Then the mass of fluid in this region is

$$M(t) = \int_{z_0}^{z_1} u(z, t)\, dz, \qquad \frac{dM}{dt} = \int_{z_0}^{z_1} u_t(z, t)\, dz,$$

The new flow rate into the section is determined by both Fick's Law and the motion of the particles. By Fick's Law, the concentration gradient contributes a term $ku_z(z_1, t) - ku_z(z_0, t)$, while the motion of the fluid contributes $Vu(z_1, t) - Vu(z_0, t)$. Thus

$$\int_{z_0}^{z_1} u_t(z, t)\, dz = ku_z(z_1, t) - ku_z(z_0, t) + Vu(z_1, t) - Vu(z_0, t).$$

Differentiating with respect to z_1 gives $u_t = ku_{zz} + Vu_z$.

1.3.9. First, we see that

$$\iiint_D \nabla \cdot \mathbf{F}\, dx = \iiint_D (r^2 x)_x + (r^2 y)_y + (r^2 z)_z\, d\mathbf{x}$$

$$= 5 \iiint_D x^2 + y^2 + z^2\, dx\, dy\, dz$$

$$= 5 \int_0^\pi \int_0^{2\pi} \int_0^a r^4 \sin\theta\, dr\, d\phi\, d\theta$$

$$= a^5 \int_0^\pi \int_0^{2\pi} \sin\theta\, d\phi\, d\theta$$

$$= 2\pi a^5 \int_0^\pi \sin\theta\, d\theta$$

$$= 4\pi a^5.$$

Next, using the fact that $\mathbf{n} = \frac{\mathbf{x}}{a}$ for $x \in$ bdy D, we see that

$$\iint_{\text{bdy } D} \mathbf{F} \cdot \mathbf{n}\, dS = \iint_{\text{bdy } D} r^2 \mathbf{x} \cdot \mathbf{n}\, dS$$

$$= \iint_{\text{bdy } D} r^2 \frac{|\mathbf{x}|^2}{a}\, dS$$

$$= \iint_{\text{bdy } D} \frac{r^4}{a}\, dS$$

$$= a^3 \iint_{\text{bdy } D} dS.$$

Now using the fact that the surface area of a sphere of radius a is $4\pi a^2$, we conclude that

$$\iint_{\text{bdy } D} \mathbf{F} \cdot \mathbf{n}\, dS = 4\pi a^5.$$

Section 1.4

1.4.1. $u(x,t) = x^2 + 2t$

1.4.5. Consider the mass $M(t)$ of material from $z = a$ up to $z = z_1$. Due to the impermeability at $z = a$, there is no loss or gain of material there. Thus

$$\int_a^{z_1} u_t dz = V u(z_1, t) + k u_z(z_1, t).$$

Putting $z_1 = a$ in this equation, we get the boundary condition $Vu + ku_z = 0$ at $z = a$.

5

1.4.6.

(a) The equilibrium will satisfy $u_{xx} = 0$ on each piece of the rod. Integrating twice gives $u_1 = c_1 x + d_1$ and $u_2 = c_2 x + d_2$. Since $u_1(0) = 0$, $d_1 = 0$, and since $u_2(L_1 + L_2) = T$ we have

$$d_2 = T - c_2(L_1 + L_2).$$

Thus the equilibrium is defined piecewise by

$$u(x) = \begin{cases} c_1 x & 0 \le x < L_1 \\ T + c_2(x - L_1 - L_2) & L_1 < x < L_1 + L_2. \end{cases}$$

To solve for c_1 and c_2 we use the fact that both u and κu_x are continuous at $x = L_1$ to get $c_1 L_1 = T - c_2 L_2$ and $\kappa_1 c_1 = \kappa_2 c_2$. Solving for c_1 and c_2 we get

$$c_1 = \frac{\kappa_2 T}{\kappa_1 L_2 + \kappa_2 L_1} \quad \text{and} \quad c_2 = \frac{\kappa_1 T}{\kappa_1 L_2 + \kappa_2 L_1}.$$

Thus

$$u(x) = \begin{cases} \dfrac{\kappa_2 T x}{\kappa_1 L_2 + \kappa_2 L_1} & 0 \le x < L_1 \\ T + \dfrac{\kappa_1 T (x - L_1 - L_2)}{\kappa_1 L_2 + \kappa_2 L_1} & L_1 < x < L_1 + L_2. \end{cases}$$

(b) When $\kappa_1 = 2$, $\kappa_2 = 1$, $L_1 = 3$, $L_2 = 2$ and $T = 10$ we have

$$u(x) = \begin{cases} \frac{10}{7} x & 0 \le x < 3 \\ 10 + \frac{20}{7}(x - 5) & 3 < x < 5. \end{cases}$$

The sketch of u is shown in Figure 2.

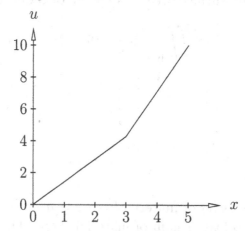

Figure 2: Solution of Exercise 1.4.6(b).

6

Section 1.5

1.5.1. Any function of the form $A \cos x + B \sin x$ solves the ODE. The condition $u(0) = 0$ implies $A = 0$, and the condition $u(L) = 0$ then implies $B \sin(L) = 0$. Thus, if L is any integer multiple of π, any function of the form $B \sin x$ is a solution, so solutions are not unique. Otherwise, if L is not an integer multiple of π, we must have $B = 0$, in which case the zero solution is the only solution.

1.5.2.

(a) Suppose u and v are both solutions. Let $w = u - v$. Then w satisfies

$$w'' + w' = 0$$
$$w'(0) = w(0) = \tfrac{1}{2}[w'(l) + w(l)].$$

Solutions of the original problem are unique if and only if $w = 0$ is the only solution of this problem. This ODE has general solution $w = C_1 + C_2 e^{-x}$. Since $w(0) = C_1 + C_2$ and $w'(0) = -C_2$, the first part of the boundary condition implies $C_1 = -2C_2$ and therefore $w = C_2(-2 + e^{-x})$. Next,

$$\frac{1}{2}[w'(l) + w(l)] = \frac{1}{2}(-C_2 e^{-l} - 2C_2 + C_2 e^{-l}) = -C_2,$$

which equals $w(0)$ and $w'(0)$. Hence any multiple of $-2 + e^{-x}$ is a solution of the above problem, so solutions of the original problem are not unique.

(b) Integrating the equation from 0 to l and using the boundary conditions gives

$$\int_0^l f(x)\, dx = \int_0^l u''(x) + u'(x)\, dx = \left[u'(x) + u(x) \right]_0^l$$
$$= u'(l) + u(l) - (u'(0) + u(0)) = 0.$$

Hence a solution can exist only if the integral of f over $[0, l]$ is zero.

Section 1.6

1.6.1.

(a) Since $a_{12}^2 - a_{11}a_{22} = 2^2 - 1 \cdot 1 = 3 > 0$, the equation is hyperbolic.

1.6.5. Let $u = v e^{\alpha x + \beta y}$. Then

$$u_x = e^{\alpha x + \beta y}(\alpha v + v_x)$$
$$u_y = e^{\alpha x + \beta y}(\beta v + v_y)$$
$$u_{xx} = e^{\alpha x + \beta y}(\alpha^2 v + 2\alpha v_x + v_{xx})$$
$$u_{yy} = e^{\alpha x + \beta y}(\beta^2 v + 2\beta v_y + v_{yy}).$$

Substituting into the equation gives

$$e^{\alpha x + \beta y}(v_{xx} + 3v_{yy} + (2\alpha - 2)v_x + (6\beta + 24)v_y + (\alpha^2 + 3\beta^2 - 2\alpha + 24\beta + 5)v) = 0.$$

To eliminate the v_x and v_y terms, let $\alpha = 1$ and $\beta = -4$. Then the new equation is

$$v_{xx} + 3v_{yy} - 44v = 0.$$

Letting $y' = \frac{1}{\sqrt{3}}y$ it follows that $v_{y'y'} = 3v_{yy}$, so $v_{xx} + v_{y'y'} - 44v = 0.$

8

Chapter 2

Section 2.1

2.1.1. By d'Alembert's formula (2.1.8),

$$u(x,t) = \frac{1}{2}\left[e^{x+ct} + e^{x-ct}\right] + \frac{1}{2c}\int_{x-ct}^{x+ct}\sin s\,ds$$
$$= \frac{1}{2}e^{x}\left(e^{ct} + e^{-ct}\right) - \frac{1}{2c}\left(\cos(x+ct) - \cos(x-ct)\right)$$
$$= e^{x}\cosh(ct) + \frac{1}{c}\sin(x)\sin(ct).$$

2.1.3. The speed of the wave is $c = \sqrt{T/\rho}$ and the left hand edge of the disturbance begins at $l/2 - a$. So the time it takes for the disturbance to travel to the position of the flea, $l/4$, is

$$\frac{(l/2 - a - l/4)}{c} = \sqrt{\rho/T}(l/4 - a).$$

2.1.6. If $ct < a$, then for any x the length of the interval $(x - ct, x + ct)$ is less than that of the interval $(-a, a)$, and therefore the intersection of these intervals has length at most $2ct$. Thus $u(x,t) \leq \frac{2ct}{2c} = t$ for all x. But at $x = 0$ the interval $(x - ct, x + ct) = (-ct, ct)$ is a subset of $(-a, a)$, so the length of the intersection is exactly $2ct$. Hence $u(0,t) = t = \max_x u(x,t)$.

On the other hand, if $ct \geq a$, then the length of the intersection is at most $2a$, so $u(x,t) \leq \frac{a}{c}$ for all x. Since this maximum length is attained when $x = 0$, it follows that $u(0,t) = \frac{a}{c} = \max_x u(x,t)$. So

$$\max_x u(x,t) = \begin{cases} t & t < \frac{a}{c} \\ \frac{a}{c} & t \geq \frac{a}{c}. \end{cases}$$

2.1.8.

(a) Applying the product rule gives $v_r = ru_r + u$ and $v_{rr} = ru_{rr} + 2u_r$. Thus

$$v_{tt} = ru_{tt} = c^2\left(ru_{rr} + 2u_r\right) = c^2 v_{rr}.$$

2.1.9. Factoring the operator yields $(\partial_x + \partial_t)(\partial_x - 4\partial_t)u = 0$. Set $v = u_x - 4u_t$. Then $v_x + v_t = 0$, so $v = h(x - t)$ and $u_x - 4u_t = h(x - t)$. One solution of this equation is $f(x - t)$, where $f'(s) = \frac{1}{5}h(s)$. The general solution of the homogeneous equation $u_x - 4u_t = 0$ is $g(4x + t)$, so

$$u(x,t) = f(x - t) + g(4x + t).$$

9

The initial conditions imply $f(x) + g(4x) = x^2$ and $-f'(x) + g'(4x) = e^x$. Differentiating the first equation leads to the system

$$\begin{array}{rcl} f'(x) & + & 4g'(4x) & = & 2x \\ -f'(x) & + & g'(4x) & = & e^x. \end{array}$$

Solving this gives $f'(x) = \frac{2}{5}x - \frac{4}{5}e^x$ and $g'(4x) = \frac{2}{5}x + \frac{1}{5}e^x$. Thus $f(x) = \frac{1}{5}x^2 - \frac{4}{5}e^x$, and $g'(s) = \frac{1}{10}s + \frac{1}{5}e^{s/4}$, so $g(s) = \frac{1}{20}s^2 + \frac{4}{5}e^{s/4}$. The solution is therefore

$$u(x,t) = \frac{1}{5}(x-t)^2 - \frac{4}{5}e^{x-t} + \frac{1}{20}(4x+t)^2 + \frac{4}{5}e^{x+t/4}$$

$$= \frac{4}{5}\left[e^{x+t/4} - e^{x-t} \right] + x^2 + \frac{1}{4}t^2.$$

Section 2.2

2.2.3.

(a) Let $v(x,t) = u(x - y, t)$. Then

$$v_{tt}(x,t) = u_{tt}(x-y,t) = c^2 u_{xx}(x-y,t) = c^2 v_{xx}(x,t).$$

2.2.5. Substituting $u_{tt} = c^2 u_{xx} - ru_t = \frac{T}{\rho}u_{xx} - ru_t$ gives

$$\frac{dE}{dt} = \int_{-\infty}^{\infty} \rho u_t u_{tt} + T u_x u_{xt} \, dx$$

$$= \int_{-\infty}^{\infty} T u_t u_{xx} + T u_x u_{xt} - \rho r u_t^2 \, dx$$

$$= \int_{-\infty}^{\infty} T(u_t u_x)_x - \rho r u_t^2 \, dx$$

$$= -\rho r \int_{-\infty}^{\infty} u_t^2 \, dx \leq 0.$$

Therefore the energy in non-increasing. (It is not necessarily the case that E is strictly decreasing.)

2.2.6.

(a) Writing $u(r,t) = \alpha(r)f(t - \beta(r))$, we have

$$u_{tt} = \alpha(r)f''(t - \beta(r))$$
$$u_r = \alpha'(r)f(t - \beta(r)) - \alpha(r)\beta'(r)f'(t - \beta(r))$$
$$u_{rr} = \alpha''(r)f(t - \beta(r)) - 2\alpha'(r)\beta'(r)f'(t - \beta(r))$$
$$- \alpha(r)\beta''(r)f'(t - \beta(r)) + \alpha(r)[\beta'(r)]^2 f''(t - \beta(r)).$$

10

Thus

$$\alpha f'' = c^2 \left(\alpha'' f - 2\alpha' \beta' f' - \alpha \beta'' f' + \alpha[\beta']^2 f'' + \frac{n-1}{r}(\alpha' f - \alpha \beta' f') \right).$$

Section 2.3

2.3.1. The maximum is $u(0,0) = 1$ and the minimum is $u(1,T) = -2kT$.

2.3.4.

(a) Since u equals zero on the lateral sides and the maximum value at time $t = 0$ is $u(1/2, 0) = 1$, the strong maximum principle implies that $u(x,t) < 1$ for all $t > 0$ and $0 < x < 1$. Since the minimum value at time $t = 0$ is $u(0,0) = 0$, the strong minimum principle implies that $u(x,t) > 0$ for all $t > 0$ and $0 < x < 1$.

(b) Let $v(x,t) = u(1-x,t)$. Then $v_x(x,t) = -u_x(1-x,t)$, so

$$v_{xx}(x,t) = u_{xx}(1-x,t) = u_t(1-x,t) = v_t(x,t),$$

and thus v is also a solution of the wave equation. Since $v(0,t) = u(1,t) = 0$, $v(1,t) = u(0,t) = 0$ and $v(x,0) = u(1-x,0) = 4(1-x)(1-(1-x)) = 4x(1-x)$, v is a solution of the diffusion equation with the same initial data and boundary data as u. Since solutions are unique, it follows that $v(x,t) = u(x,t)$ for all $t \geq 0$ and $0 \leq x \leq 1$.

(c) Since u vanishes as $x = 0$ and $x = 1$ for all t, we have

$$\begin{aligned}
\frac{d}{dt} \int_0^1 u^2(x,t)\,dx &= 2 \int_0^1 u(x,t) u_t(x,t)\,dx \\
&= 2 \int_0^1 u(x,t) u_{xx}(x,t)\,dx \\
&= 2u(x,t)u_x(x,t) \Big|_{x=0}^{x=1} - 2 \int_0^1 u_x^2(x,t)\,dx \\
&= -2 \int_0^1 u_x^2(x,t)\,dx.
\end{aligned}$$

The integral cannot be zero since this would imply $u_x(x,t) = 0$, which means that for each t, $u(x,t)$ would be constant in x. Since $u(0,t) = 0$ that constant would be zero. But by part (a), u is positive for $0 < x < 1$. Thus $\int_0^1 u^2(x,t)\,dx$ is strictly decreasing, since its derivative is negative.

11

2.3.5. The PDE is $u_t = xu_{xx}$.

(a) Let $u = -2xt - x^2$. It is elementary to check that it is a solution. We use calculus to find its maximum and minimum in the rectangle R. To do this, we check the interior using $u_t = u_x = 0$ to get the point $(0,0)$ where $u = 0$. Then we check each of the four sides by taking one derivative in each case, and finally we consider the corners separately. Among all these points that are in the rectangle R, we look for the largest and the smallest values of u. We find that the maximum value is $u = 1$ at the point $(-1, 1)$, which is on the top of R, and the minimum value is $u = -8$ at $(2, 1)$, which is a corner point. The maximum occurs on the top of R, which violates the (diffusion) maximum principle.

(b) At any maximum point on the top, we must have $u_t \geq 0$ and $u_{xx} \leq 0$. At our maximum point $(-1, 1)$, $u_t = 2 > 0$ and $xu_{xx} = (-1)(-2) = 2 > 0$. There is no contradiction to the PDE $u_t = xu_{xx}$, even if the inequalities were strict. (The purpose of introducing ϵx^2 in the text was to make the inequalities strict, but here this would not help.)

Section 2.4

2.4.1. Letting $p = (y - x)/\sqrt{4kt}$ in equation (2.4.8) gives

$$
\begin{aligned}
u(x,t) &= \frac{1}{\sqrt{4\pi kt}} \int_{-l}^{l} e^{-(y-x)^2/4kt}\, dy \\
&= \frac{1}{\sqrt{\pi}} \int_{(-l-x)/\sqrt{4kt}}^{(l-x)/\sqrt{4kt}} e^{-p^2}\, dp \\
&= \frac{1}{\sqrt{\pi}} \int_{0}^{(l-x)/\sqrt{4kt}} e^{-p^2}\, dp - \frac{1}{\sqrt{\pi}} \int_{0}^{(-l-x)/\sqrt{4kt}} e^{-p^2}\, dp \\
&= \frac{1}{2}\mathscr{E}\mathrm{rf}\left(\frac{l-x}{\sqrt{4kt}}\right) - \frac{1}{2}\mathscr{E}\mathrm{rf}\left(\frac{-l-x}{\sqrt{4kt}}\right).
\end{aligned}
$$

2.4.3. Using equation (2.4.8), we have

$$
u(x,t) = \frac{1}{\sqrt{4\pi kt}} \int_{-\infty}^{\infty} e^{-(x-y)^2/4kt} e^{3y}\, dy.
$$

Completing the square, the exponent becomes

$$
\frac{-y^2 + 2xy + 12kty - x^2}{4kt} = \frac{-(y - x - 6kt)^2 + 36k^2t^2 + 12kxt}{4kt}.
$$

So letting $p = (y - x - 6kt)/\sqrt{4kt}$, we get

$$
u(x,t) = \frac{1}{\sqrt{\pi}} \int_{-\infty}^{\infty} e^{-p^2} e^{9kt+3x}\, dp = e^{9kt+3x}.
$$

2.4.6. Let $I = \int_0^\infty e^{-x^2}\, dx$. Then $I = \int_0^\infty e^{-y^2}\, dy$, so

$$
\begin{aligned}
I^2 &= \int_0^\infty e^{-x^2}\, dx \int_0^\infty e^{-y^2}\, dy \\
&= \int_0^\infty \left(\int_0^\infty e^{-x^2}\, dx \right) e^{-y^2}\, dy \\
&= \int_0^\infty \int_0^\infty e^{-x^2} e^{-y^2}\, dx\, dy \\
&= \int_0^\infty \int_0^\infty e^{-(x^2+y^2)}\, dx\, dy.
\end{aligned}
$$

Changing to polar coordinates, this becomes

$$
I^2 = \int_0^{\pi/2} \int_0^\infty e^{-r^2} r\, dr\, d\theta = \int_0^{\pi/2} -\frac{1}{2} e^{-r^2} \Big|_0^\infty d\theta = \int_0^{\pi/2} \frac{1}{2}\, d\theta = \frac{\pi}{4}.
$$

Thus $I = \frac{\sqrt{\pi}}{2}$.

2.4.8. For each fixed t, $S(x,t)$ is a decreasing function of $|x|$, so

$$
\max_{\delta \le |x| < \infty} S(x,t) = S(\delta, t) = \frac{1}{\sqrt{4\pi kt}} e^{-\delta^2/4kt}.
$$

Now let $t = 1/4ks$. Then, by L'Hopital's rule,

$$
\lim_{t \to 0} \frac{1}{\sqrt{4\pi kt}} e^{-\delta^2/4kt} = \lim_{s \to \infty} \frac{1}{\sqrt{\pi}} \frac{\sqrt{s}}{e^{\delta^2 s}} = \lim_{s \to \infty} \frac{1}{\sqrt{\pi}} \frac{1}{2\sqrt{s}\delta^2 e^{\delta^2 s}} = 0.
$$

2.4.9. Let u be the solution. Then $(u_{xxx})_t = (u_t)_{xxx} = (ku_{xx})_{xxx} = k(u_{xxx})_{xx}$, so u_{xxx} is a solution, and $u_{xxx}(x,0) = \frac{d^3}{dx^3}(x^2) = 0$. Hence u_{xxx} is the zero function. Integrating three times with respect to x gives $u(x,t) = A(t)x^2 + B(t)x + C(t)$. Substituting u into the diffusion equation $u_t = ku_{xx}$ gives

$$
A'(t)x^2 + B'(t)x + C'(t) = 2kA(t).
$$

Equating coefficients gives $A' = B' = 0$ and $C' = 2kA$. Thus A and B are constant and $C = 2kAt + D$, so $u(x,t) = Ax^2 + Bx + 2kAt + D$. Now the initial condition $u(x,t) = x^2$ implies that $A = 1$ and $B = D = 0$, so $u(x,t) = x^2 + 2kt$.

2.4.12.

(a) $Q(x,t) = \frac{1}{2} + \frac{1}{2}\mathscr{E}\mathrm{rf}\left(\frac{x}{\sqrt{4kt}} \right)$.

(b) Since the Taylor series for e^z centered at $z = 0$ is $\sum_{j=0}^\infty \frac{z^j}{j!}$, it follows that

$$
e^{-p^2} = \sum_{j=0}^\infty \frac{(-1)^j p^{2j}}{j!}
$$

13

and therefore

$$\mathscr{E}\mathrm{rf}(x) = \frac{2}{\sqrt{\pi}} \sum_{j=0}^{\infty} \frac{(-1)^j x^{2j+1}}{(2j+1)j!}.$$

(c) Using the $j = 0$ and $j = 1$ terms above yields

$$Q(x,t) = \frac{1}{2} + \frac{1}{\sqrt{\pi}} \left(\frac{x}{\sqrt{4kt}} - \frac{1}{3} \left(\frac{x}{\sqrt{4kt}} \right)^3 \right).$$

(d) By Taylor's Theorem, the remainder takes the form

$$\frac{1}{\sqrt{\pi}} \frac{(z/\sqrt{4kt})^5}{10}$$

for some z between 0 and x. This is small for x fixed and t large.

2.4.17. Let $u = e^{-bt^3/3}v$. Then $u_t = e^{-bt^3/3}(-bt^2 v + v_t)$ and $u_{xx} = e^{-bt^3/3}v_{xx}$, so substituting into the equation $u_t - ku_{xx} + bt^2 u = 0$ gives $v_t - kv_{xx} = 0$. When $t = 0$, $u = v$ so the initial data is unchanged. So v is given by equation (2.4.8) and thus

$$u(x,t) = \frac{e^{-bt^3/3}}{\sqrt{4\pi kt}} \int_{-\infty}^{\infty} e^{-(x-y)^2/4kt} \phi(y)\,dy.$$

2.4.18. Let $v(y,t) = u(y + Vt, t)$. Then $v_t(y,t) = Vu_x(y + Vt, t) + u_t(y + Vt, t)$, $v_y(y,t) = u_x(y + Vt, t)$ and $v_{yy}(y,t) = u_{xx}(y + Vt, t)$. Thus

$$v_t(y,t) - kv_{yy}(y,t) = u_t(y + Vt, t) - ku_{xx}(y + Vt, t) + Vu_x(y + Vt, t) = 0,$$

so v is a solution of the diffusion equation, with initial data $v(y,0) = u(y,0) = \phi(y)$. Thus, by equation (2.4.6),

$$v(y,t) = \int_{-\infty}^{\infty} S(y - w, t)\phi(w)\,dw,$$

which means

$$u(x,t) = v(x - Vt, t) = \int_{-\infty}^{\infty} S(x - Vt - w, t)\phi(w)\,dw.$$

Section 2.5

2.5.1. Consider for instance the solution of the wave equation in one dimension with initial data $\phi(x) = 0$ and

$$\psi(x) = \begin{cases} 1 & |x| < 1 \\ 0 & |x| \geq 1 \end{cases},$$

which corresponds to the "hammer blow". Since there is no boundary, the maximum principle would state that the maximum is attained initially. But the solution is zero initially, and takes on positive values for $t > 0$. Thus the maximum principle does not hold for the wave equation.

14

Chapter 3

Section 3.1

3.1.2. Introduce the function $v(x,t)$ such that $v(x,t) = u(x,t) - 1$. Then v satisfies

$$
\begin{aligned}
v_t - kv_{xx} &= 0 & 0 < x < \infty \\
v(x,0) &= -1 & 0 < x < \infty \\
v(0,t) &= 0.
\end{aligned}
$$

We know by equation (3.1.6) that v is given by

$$
v(x,t) = -\frac{1}{\sqrt{4\pi kt}} \int_0^\infty \left[e^{-(x-y)^2/4kt} - e^{-(x+y)^2/4kt} \right] dy.
$$

Making the change of variables $p = (y-x)/\sqrt{4kt}$, we see that

$$
\frac{1}{\sqrt{4\pi kt}} \int_0^\infty e^{-(x-y)^2/4kt} \, dy = \frac{1}{\sqrt{\pi}} \int_{-x/\sqrt{4kt}}^\infty e^{-p^2} \, dp.
$$

Similarly, making the change of variables $q = (y+x)/\sqrt{4kt}$, we see that

$$
\frac{1}{\sqrt{4\pi kt}} \int_0^\infty e^{-(x+y)^2/4kt} \, dy = \frac{1}{\sqrt{\pi}} \int_{x/\sqrt{4kt}}^\infty e^{-q^2} \, dq.
$$

Using these facts, we conclude that

$$
\begin{aligned}
v(x,t) &= -\frac{1}{\sqrt{\pi}} \int_{-x/\sqrt{4kt}}^{x/\sqrt{4kt}} e^{-p^2} \, dp \\
&= -\frac{2}{\sqrt{\pi}} \int_0^{x/\sqrt{4kt}} e^{-p^2} \, dp \\
&= -\mathscr{E}\mathrm{rf}(x/\sqrt{4kt}).
\end{aligned}
$$

We conclude that $u(x,t) \equiv v(x,t) + 1$ for $x > 0$ satisfies the desired problem. In particular, we conclude that $u(x,t) = 1 - \mathscr{E}\mathrm{rf}(x/\sqrt{4kt})$.

3.1.3. Let $\phi_{\text{even}}(x)$ be the even extension of $\phi(x)$ to the whole real line. Let $v(x,t)$ be the solution of

$$
\begin{cases}
v_t - kv_{xx} = 0 & -\infty < x < \infty, t > 0 \\
v(x,0) = \phi_{\text{even}}(x).
\end{cases}
$$

Then $w(x,t) \equiv v(x,t)$ for $x > 0$ will be a solution of the heat equation on the half-line with

Neumann boundary conditions. We see that

$$v(x,t) = \int_{-\infty}^{\infty} S(x-y,t)\phi_{\text{even}}(y)\,dy$$

$$= \frac{1}{\sqrt{4\pi kt}} \int_{-\infty}^{\infty} e^{-(x-y)^2/4kt} \phi_{\text{even}}(y)\,dy$$

$$= \frac{1}{\sqrt{4\pi kt}} \int_{0}^{\infty} e^{-(x-y)^2/4kt} \phi(y)\,dy + \frac{1}{\sqrt{4\pi kt}} \int_{-\infty}^{0} e^{-(x-y)^2/4kt} \phi(-y)\,dy.$$

Making the change of variables $z = -y$ in the second integral, we see that

$$\frac{1}{\sqrt{4\pi kt}} \int_{-\infty}^{0} e^{-(x-y)^2/4kt} \phi(-y)\,dy = -\frac{1}{\sqrt{4\pi kt}} \int_{\infty}^{0} e^{-(x+z)^2/4kt} \phi(z)\,dz$$

$$= \frac{1}{\sqrt{4\pi kt}} \int_{0}^{\infty} e^{-(x+y)^2/4kt} \phi(y)\,dy.$$

Combining this term with the first integral above, we get the desired result.

Section 3.2

3.2.2. With the Neumann BC we take the even extension of $\psi(x)$. Thus

$$u(x,t) = \frac{1}{2c} \int_{x-ct}^{x+ct} \psi_{\text{even}}(y)\,dy = \frac{V}{2c} \cdot \text{length of} \{[-2a,-a] \cup [a,2a]\} \cap [x-ct, x+ct].$$

At $t = 0$, $u(x,t) \equiv 0$. At $t = a/c$,

$$u\left(x, \frac{a}{c}\right) = \frac{1}{2c} \int_{x-a}^{x+a} \psi_{\text{even}}(y)\,dy.$$

Therefore,

$$u(0, a/c) = \frac{1}{2c} \int_{-a}^{a} \psi_{\text{even}}(y)\,dy = 0$$

$$u(a, a/c) = \frac{1}{2c} \int_{0}^{2a} \psi_{\text{even}}(y)\,dy = (V/2c) \cdot \text{ length of } [a, 2a] = Va/2c$$

$$u(2a, a/c) = \frac{1}{2c} \int_{a}^{3a} \psi_{\text{even}}(y)\,dy = Va/2c$$

$$u(3a, a/c) = \frac{1}{2c} \int_{2a}^{4a} \psi_{\text{even}}(y)\,dy = 0.$$

Thus the graph of $u(x, a/c)$ is a straight line from $(0,0)$ to $(a, Va/2c)$, then constant to $(2a, Va/2c)$, then a straight line to $(3a, 0)$, and then identically 0 for $x > 3a$. See Figure 3. Similarly, we consider the other values of t. We find that $u(x, 3a/2c)$ equals $Va/2c$ for

16

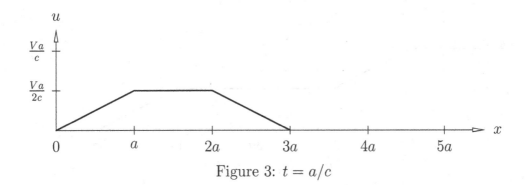

Figure 3: $t = a/c$

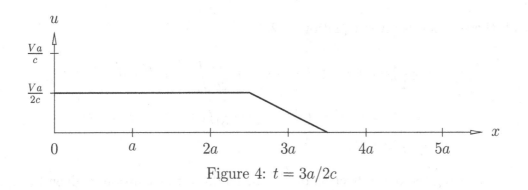

Figure 4: $t = 3a/2c$

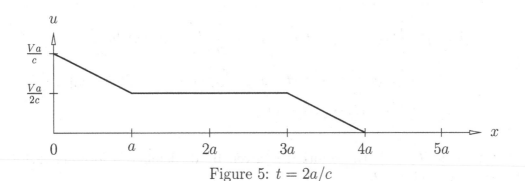

Figure 5: $t = 2a/c$

$0 \leq x \leq 5a/2$, then drops linearly to 0 and vanishes for $x \geq 7a/2$. See Figure 4 Likewise, $u(x, 2a/c)$ equals Va/c for $x = 0$, drops linearly to $Va/2c$ at $x = a$, equals $Va/2c$ for $a \leq x \leq 3a$, then drops linearly to 0 at $x = 4a$ and vanishes for $x \geq 4a$. See Figure 5. Finally, $u(x, 3a/c)$ equals Va/c for $0 \leq x \leq a$, drops linearly to $Va/2c$ at $x = 2a$, equals $Va/2c$ for $2a \leq x \leq 4a$, then drops linearly to 0 at $x = 5a$ and vanishes for $x \geq 5a$. See Figure 6.

3.2.3. The problem that u satisfies is given by

$$\begin{cases} u_{tt} - c^2 u_{xx} = 0 & 0 < x < \infty, t > 0 \\ u(x, 0) = f(x) & 0 < x < \infty \\ u_t(x, 0) = cf'(x) & 0 < x < \infty \\ u(0, t) = 0. \end{cases}$$

17

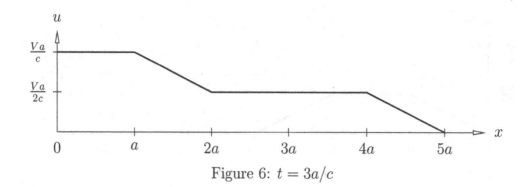

Figure 6: $t = 3a/c$

For $x > ct$, the solution is given equation (3.2.2),

$$u(x,t) = \frac{1}{2}[f(x+ct) + f(x-ct)] + \frac{1}{2c}\int_{x-ct}^{x+ct} cf'(y)\, dy$$
$$= \frac{1}{2}[f(x+ct) + f(x-ct)] + \frac{1}{2}[f(x+ct) - f(x-ct)]$$
$$= f(x+ct).$$

For $0 < x < ct$, using the odd reflection of our initial data, the solution is given by equation (3.2.3),

$$u(x,t) = \frac{1}{2}[f(x+ct) - f(ct-x)] + \frac{1}{2c}\int_{ct-x}^{x+ct} cf'(y)\, dy$$
$$= \frac{1}{2}[f(x+ct) - f(ct-x)] + \frac{1}{2}[f(x+ct) - f(ct-x)]$$
$$= f(x+ct) - f(ct-x).$$

3.2.5. For $x > 2t$ $(t > 0)$, the solution is given by d'Alembert's formula (2.1.8), with $\phi(x) \equiv 1$, $\psi(x) \equiv 0$, and $c = 2$,

$$u(x,t) = 1 \qquad \text{for } x > 2t.$$

For $x < 2t$, we extend the initial data to be odd with respect to the $x = 0$ line. In this case, our solution is

$$u(x,t) = \frac{1}{2}[1 - 1] = 0 \qquad \text{for } x < 2t.$$

We see that the solution has a singularity along the line $x = 2t$.

3.2.6. From equation (2.1.3), $u = f(x+ct) + g(x-ct)$ for $x > 0$ and $t \geq 0$. Thus $u_t = cf'(x+ct) - cg'(x-ct)$. Putting $t = 0$ gives

$$0 = f(x) + g(x)$$
$$V = cf'(x) - cg'(x)$$

18

for $x > 0$. Thus $0 = f'(x) + g'(x)$ and we easily solve to find $f'(x) = \frac{V}{2c} = -g'(x)$. Thus $f(x) = \frac{V}{2c}x$ and $g(x) = -\frac{V}{2c}x$ for $x > 0$.

On the other hand, the boundary condition yields

$$0 = u_t(0,t) + au_x(0,t) = [cf'(ct) - cg'(-ct)] + a[f'(ct) + g'(-ct)]$$
$$= (c+a)f'(ct) + (-c+a)g'(-ct),$$

so that $g'(-ct) = -\frac{a+c}{a-c}f'(ct)$ for $t > 0$. In other words, $g'(-y) = -\frac{a+c}{a-c}f'(y)$ for $y > 0$. Therefore $g(-y) = \frac{a+c}{a-c}f(y) + k$ for $y > 0$, for some constant k. Thus $g(x) = -\frac{a+c}{a-c}\frac{V}{2c}x + k$, and since we want g to be continuous at $x = 0$ we take $k = 0$. Therefore

$$u(x,t) = \begin{cases} Vt & x > ct \\ \frac{V}{c-a}(x - at) & x < ct. \end{cases}$$

Note: At the corner $(0,0)$ the initial condition leads to $u_t + au_x = V + a \cdot 0 = V$, which contradicts the boundary condition. Therefore, although the problem is solvable, there is a discontinuity in the solution (along the line $x = ct$).

3.2.9.

(a) The solution is given by d'Alembert's formula (2.1.8) as

$$u(x,t) = \frac{1}{2}[\phi_{\text{ext}}(x+t) + \phi_{\text{ext}}(x-t)] + \frac{1}{2}\int_{x-t}^{x+t}\psi_{\text{ext}}(y)\,dy,$$

where ϕ_{ext} and ψ_{ext} are the odd, period 2 extensions of the initial data.

Therefore,

$$\phi_{\text{ext}}(2/3 + 2) = \phi\left(\frac{2}{3}\right) = \frac{4}{27}.$$

and

$$\phi_{\text{ext}}(2/3 - 2) = \phi\left(\frac{2}{3}\right) = \frac{4}{27}.$$

Further, since ψ_{ext} is odd and has period 2,

$$\int_{2/3-2}^{2/3+2}\psi_{\text{ext}}(y)\,dy = 0,$$

and we conclude that

$$u(x,t) = \frac{1}{2}[\phi_{\text{ext}}(x+t) + \phi_{\text{ext}}(x-t)] = \frac{4}{27}.$$

(b) Using the same technique as above, it follows that

$$\phi_{\text{ext}}(1/4 + 7/2) = \phi_{\text{ext}}(15/4) = \phi_{\text{ext}}(-1/4) = -\phi\left(\frac{1}{4}\right) = -\frac{3}{64}$$

and

$$\phi_{\text{ext}}(1/4 - 7/2) = \phi_{\text{ext}}(-13/4) = \phi_{\text{ext}}(3/4) = \phi\left(\frac{3}{4}\right) = \frac{9}{64}.$$

Again, using the fact that ψ_{ext} is odd and has period 2, we have

$$\int_{1/4-7/2}^{1/4+7/2} \psi_{\text{ext}}(y)\,dy = \int_{-13/4}^{15/4} \psi_{\text{ext}}(y)\,dy = \int_{13/4}^{15/4} \psi_{\text{ext}}(y)\,dy = \int_{-3/4}^{-1/4} \psi_{\text{ext}}(y)\,dy$$

$$= \int_{-3/4}^{-1/4} -\psi(-y)\,dy = -\int_{1/4}^{3/4} \psi(y)\,dy = -\frac{13}{96}.$$

Therefore, we conclude that

$$u(x,t) = \frac{1}{2}[\phi_{\text{ext}}(x+t) + \phi_{\text{ext}}(x-t)] + \frac{1}{2}\int_{x-t}^{x+t} \psi_{\text{ext}}(y)\,dy$$

$$= \frac{1}{2}\left[-\frac{3}{64} + \frac{9}{64}\right] + \frac{1}{2}\left[-\frac{13}{96}\right] = -\frac{1}{48}.$$

Section 3.3

3.3.1. Let $\phi_{\text{odd}}(x)$ and $f_{\text{odd}}(x,t)$ be the odd extensions of ϕ and f with respect to $x = 0$ respectively. Let v be the solution of

$$\begin{cases} v_t - kv_{xx} = f_{\text{odd}} & -\infty < x < \infty, t > 0 \\ v(x,0) = \phi_{\text{odd}} & -\infty < x < \infty. \end{cases}$$

Then, by equation (3.3.2),

$$v(x,t) = \int_{-\infty}^{\infty} S(x-y,t)\phi_{\text{odd}}(y)\,dy + \int_0^t \int_{-\infty}^{\infty} S(x-y,t-s)f_{\text{odd}}(y,s)\,dy\,ds,$$

where $S(y,t) = \frac{1}{\sqrt{4\pi kt}}e^{-y^2/4kt}$. We note that $v(0,t) = 0$ because $S(y,t)$ is even with respect to $y = 0$. Therefore, letting $u(x,t) = v(x,t)$ for $x > 0$, we see that u is the solution of the desired problem. Splitting up the integrals at $y = 0$ and substituting $z = -y$ in the integral over $(-\infty, 0)$, the solution becomes

$$u(x,t) = \int_0^{\infty} [S(x-y,t) - S(x+y,t)]\,\phi(y)\,dy$$

$$+ \int_0^t \int_{-\infty}^{\infty} [S(x-y,t-s) - S(x+y,t-s)]\,f(y,s)\,dy\,ds.$$

20

Section 3.4

3.4.1. By equation (3.4.3), the solution is given by

$$u(x,t) = \frac{1}{2c} \int_0^t \int_{x-c(t-s)}^{x+c(t-s)} f(y,s)\, dy\, ds = \frac{1}{2c} \int_0^t \int_{x-c(t-s)}^{x+c(t-s)} ys\, dy\, ds$$

$$= \frac{1}{2c} \int_0^t s \left. \frac{y^2}{2} \right|_{x-c(t-s)}^{x+c(t-s)} ds = \frac{1}{2c} \int_0^t s \left[\frac{(x+c(t-s))^2}{2} - \frac{(x-c(t-s))^2}{2} \right] ds$$

$$= \frac{1}{4c} \int_0^t s[4cxt - 4xcs]\, ds = \int_0^t [xts - xs^2]\, ds$$

$$= \left. \frac{xts^2}{2} - \frac{xs^3}{3} \right|_{s=0}^{s=t} = \frac{xt^3}{6}.$$

3.4.3.

$$u(x,t) = \frac{1}{2}[\sin(x+ct) + \sin(x-ct)] + \frac{1}{2c} \int_{x-ct}^{x+ct} (1+y)\, dy$$

$$+ \frac{1}{2c} \int_0^t \int_{x-c(t-s)}^{x+c(t-s)} \cos y\, dy\, ds.$$

Now the first integral is evaluated as follows.

$$\frac{1}{2c} \int_{x-ct}^{x+ct} (1+y)\, dy = \frac{1}{2c} \left. \left[y + \frac{y^2}{2} \right] \right|_{x-ct}^{x+ct}$$

$$= \frac{1}{2c} \left[(x+ct) + \frac{(x+ct)^2}{2} - (x-ct) - \frac{(x-ct)^2}{2} \right]$$

$$= \frac{1}{2c}[2ct + 2xct]$$

$$= t + xt.$$

The second integral is evaluated as follows.

$$\frac{1}{2c} \int_0^t \int_{x-c(t-s)}^{x+c(t-s)} \cos y\, dy\, ds = \frac{1}{2c} \int_0^t \left. \sin y \right|_{x-c(t-s)}^{x+c(t-s)} ds$$

$$= \frac{1}{2c} \int_0^t [\sin(x+c(t-s)) - \sin(x-c(t-s))]\, ds$$

$$= \left. \frac{1}{2c^2}[\cos(x+c(t-s)) + \cos(x-c(t-s))] \right|_{s=0}^{s=t}$$

$$= \frac{1}{c^2}\cos(x) - \frac{1}{2c^2}[\cos(x+ct) + \cos(x-ct)].$$

Therefore, the solution is given by

$$u(x,t) = \frac{1}{2}[\sin(x+ct) + \sin(x-ct)] + t + xt + \frac{1}{c^2}\cos(x) - \frac{1}{2c^2}[\cos(x+ct) + \cos(x-ct)]$$

$$= \sin x \cos ct + t(x+1) + \frac{1}{c^2}\cos x[1 - \cos ct].$$

21

3.4.6.

(a) The inhomogeneous wave equation factors as

$$(\partial_t - c\partial_x)(\partial_t + c\partial_x)u = f.$$

Now if $v = u_t + cu_x$, then $v_t - cv_x = f$.

(b) Introduce the new variables

$$t' = t + cx$$
$$x' = ct - x.$$

With respect to this change of variables,

$$x = \frac{1}{1+c^2}(ct' - x')$$
$$t = \frac{1}{1+c^2}(t' + cx').$$

By the chain rule,

$$u_{t'} = u_x \frac{\partial x}{\partial t'} + u_t \frac{\partial t}{\partial t'} = u_x\left(\frac{c}{1+c^2}\right) + u_t\left(\frac{1}{1+c^2}\right) = \frac{1}{1+c^2}(u_t + cu_x) = \frac{1}{1+c^2}v(x,t).$$

Therefore, defining

$$\widetilde{u}(x',t') = u(x(x',t'),t(x',t'))$$
$$\widetilde{v}(x',t') = v(x(x',t'),t(x',t')),$$

we see that

$$\frac{\partial \widetilde{u}}{\partial t'}(x',t') = \frac{1}{1+c^2}\widetilde{v}(x',t').$$

Integrating with respect to t', we have

$$u(x,t) = \widetilde{u}(x',t')$$

$$= \frac{1}{1+c^2}\int^{t'} \widetilde{v}(x',s)\,ds$$

$$= \frac{1}{1+c^2}\int^{t+cx} v(x(x',s),t(x',s))\,ds$$

$$= \frac{1}{1+c^2}\int^{t+cx} v\left(\frac{1}{1+c^2}(cs - x'), \frac{1}{1+c^2}(s + cx')\right)\,ds.$$

Making the change of variables $\widetilde{s} = \frac{1}{1+c^2}(s + cx')$, we see that

$$u(x,t) = \int^{\frac{1}{1+c^2}(t+cx+cx')} v(c\widetilde{s} - x', \widetilde{s})\,d\widetilde{s}$$

$$= \int^{t} v(x - ct + c\widetilde{s}, \widetilde{s})\,d\widetilde{s}.$$

The lower limit can be arbitrary. Assuming $u(x,0) = 0$, we make the lower limit 0, in which case

$$u(x,t) = \int_0^t v(x - ct + cs, s)\,ds.$$

(c) Make the change of variables

$$t' = cx - t$$
$$x' = x + ct.$$

With respect to this change of variables, we have

$$t = \frac{1}{1 + c^2}(cx' - t')$$
$$x = \frac{1}{1 + c^2}(x' + ct').$$

By the chain rule, we have

$$\partial_{t'} v(x,t) = v_t \frac{\partial t}{\partial t'} + v_x \frac{\partial x}{\partial t'}$$

$$= v_t \left(-\frac{1}{1 + c^2}\right) + v_x \left(\frac{c}{1 + c^2}\right)$$

$$= -\frac{1}{1 + c^2}(v_t - cv_x)$$

$$= -\frac{1}{1 + c^2}f(x,t).$$

Letting

$$\widetilde{v}(x',t') = v(x(x',t'), t(x',t'))$$
$$\widetilde{f}(x',t') = f(x(x',t'), t(x',t')),$$

we have

$$\frac{\partial}{\partial t'}\widetilde{v} = -\frac{1}{1 + c^2}\widetilde{f}(x',t').$$

Integrating with respect to t', we have

$$v(x,t) = \widetilde{v}(x',t')$$

$$= -\frac{1}{1 + c^2}\int^{t'} \widetilde{f}(x',s)\,ds$$

$$= -\frac{1}{1 + c^2}\int^{cx-t} f(x(x',s), t(x',s))\,ds$$

$$= -\frac{1}{1 + c^2}\int^{cx-t} f\left(\frac{1}{1 + c^2}(x + ct + cs), \frac{1}{1 + c^2}(c(x + ct) - s)\right)\,ds.$$

Making the substitution $\tilde{s} = \frac{1}{1+c^2}(c(x+ct) - s)$, we have

$$v(x, t) = \int^t f(x + ct - c\tilde{s}, \tilde{s}) \, d\tilde{s}.$$

The lower limit can be arbitrary. Assuming $v(x, 0) = 0$, we make the lower limit zero. Therefore,

$$v(x, t) = \int_0^t f(x + ct - cs, s) \, ds.$$

(d) Using the results from parts (b) and (c), we see that

$$u(x, t) = \int_0^t v(x - ct + cs, s) \, ds$$
$$= \int_0^t \int_0^s f(x - ct + 2cs - c\tilde{s}, \tilde{s}) \, d\tilde{s} \, ds.$$

Making the change of variables,

$$y = x - ct + 2cs - c\tilde{s}$$
$$s' = \tilde{s},$$

we see that the triangle in the $s\tilde{s}$-plane with vertices at $(0, 0)$, $(t, 0)$ and (t, t) is transformed into the triangle in the ys'-plane with vertices at $(x - ct, 0)$, $(x + ct, 0)$ and (x, t). The Jacobian of this transformation is given by

$$J = \left| \det \begin{bmatrix} \frac{\partial s}{\partial y} & \frac{\partial s}{\partial s'} \\ \frac{\partial \tilde{s}}{\partial y} & \frac{\partial \tilde{s}}{\partial s'} \end{bmatrix} \right| = \left| \det \begin{bmatrix} \frac{1}{2c} & \frac{1}{2} \\ 0 & 1 \end{bmatrix} \right| = \frac{1}{2c}.$$

Therefore, we have

$$u(x, t) = \int_0^t \int_0^s f(x - ct + 2cs - c\tilde{s}, \tilde{s}) \, d\tilde{s} \, ds = \frac{1}{2c} \iint_\Delta f(y, s) \, dy \, ds$$

where Δ is the characteristic triangle.

3.4.8. The source operator $\mathscr{S}(t)$ is defined as the operator such that

$$\mathscr{S}(t)\psi = \frac{1}{2c} \int_{x-ct}^{x+ct} \psi(y) \, dy.$$

We note that

$$\mathscr{S}(0)\psi = \frac{1}{2c} \int_x^x \psi(y) \, dy = 0.$$

Therefore, we say $\mathscr{S}(0) = 0$ as an operator. Next, we note that

$$(\mathscr{S}(t)\psi)_t = \frac{1}{2c}[\psi(x + ct)c - \psi(x - ct)(-c)]$$
$$= \frac{1}{2}[\psi(x + ct) + \psi(x - ct)].$$

24

Therefore,

$$(\mathscr{S}(0)\psi)_t = \frac{1}{2}[\psi(x) + \psi(x)] = \psi(x).$$

We say that

$$\mathscr{S}_t(0) = I$$

as an operator. Next,

$$[\mathscr{S}(t)\psi]_{tt} = \frac{1}{2}[\psi'(x+ct)c + \psi'(x-ct)(-c)]$$
$$= \frac{c}{2}[\psi'(x+ct) - \psi'(x-ct)].$$

Finally,

$$(\mathscr{S}(t)\psi)_x = \frac{1}{2c}[\psi(x+ct) - \psi(x-ct)]$$

implies

$$(\mathscr{S}(t)\psi)_{xx} = \frac{1}{2c}[\psi'(x+ct) - \psi'(x-ct)].$$

Therefore,

$$(\mathscr{S}(t)\psi)_{tt} - c^2(\mathscr{S}(t)\psi)_{xx} = \frac{c}{2}[\psi'(x+ct) - \psi'(x-ct)] - \frac{c^2}{2c}[\psi'(x+ct) - \psi'(x-ct)] = 0.$$

Therefore,

$$\mathscr{S}_{tt} - c^2\mathscr{S}_{xx} = 0$$

as an operator.

3.4.9. For

$$u(t) = \int_0^t \mathscr{S}(t-s)f(s)\,ds,$$

we see that

$$u(0) = \int_0^0 \mathscr{S}(0-s)f(s)\,ds = 0.$$

In addition,

$$u_t(t) = \mathscr{S}(t-t)f(t) + \int_0^t \mathscr{S}_t(t-s)f(s)\,ds$$
$$= \mathscr{S}(0)f(t) + \int_0^t \mathscr{S}_t(t-s)f(s)\,ds$$
$$= \int_0^t \mathscr{S}_t(t-s)f(s)\,ds.$$

Therefore,

$$u_t(0) = \int_0^0 \mathscr{S}_t(0-s)f(s)\,ds = 0.$$

25

Using u_t calculated above, we see that

$$u_{tt} = \mathscr{S}_t(t-t)f(t) + \int_0^t \mathscr{S}_{tt}(t-s)f(s)\,ds$$

$$= f(t) + \int_0^t \mathscr{S}_{tt}(t-s)f(s)\,ds.$$

Next,

$$u_x = \int_0^t \mathscr{S}_x(t-s)f(s)\,ds$$

implies

$$u_{xx} = \int_0^t \mathscr{S}_{xx}(t-s)f(s)\,ds.$$

Combining these terms, we see that

$$u_{tt} - c^2 u_{xx} = f(t) + \int_0^t [\mathscr{S}_{tt}(t-s)f(s) - c^2\mathscr{S}_{xx}(t-s)f(s)]\,ds$$

$$= f(t) + \int_0^t [\mathscr{S}_{tt}(t-s) - c^2\mathscr{S}_{xx}(t-s)]f(s)\,ds$$

$$= f(t),$$

using the results from Exercise 3.4.8.

3.4.13. Let $v(x,t) = u(x,t) - t^2$. Then v satisfies

$$\begin{array}{ll}
v_{tt} - c^2 v_{xx} = -2 & 0 < x < \infty,\ 0 < t < \infty \\
v(x,0) = x & 0 < x < \infty \\
v_t(x,0) = 0 & 0 < x < \infty \\
v(0,t) = 0 & 0 < t < \infty.
\end{array}$$

For $x > ct$ this has solution

$$v(x,t) = x + \frac{1}{2c}\int_0^t \int_{x-c(t-s)}^{x+c(t-s)} -2\,dy\,ds = x - t^2.$$

For $x < ct$ the domain of dependence D is given as in Figure 3.2.2. The area of D is $ct^2 - (ct-x)(t-x/c) = 2tx - x^2/c$, so the solution is given by

$$v(x,t) = \frac{1}{2}[ct + x - (ct-x)] + \frac{1}{2c}\iint_D -2\,dA = x - 2tx/c + x^2/c^2.$$

Hence

$$u(x,t) = \begin{cases} x & x > ct \\ x - \dfrac{2tx}{c} + \dfrac{x^2}{c^2} + t^2 & x < ct. \end{cases}$$

3.4.14.

Solution 1. This technique is somewhat ad hoc, but it yields the solution more quickly. Take a look at the second solution for a more methodical approach.

First recall that solutions of the wave equation take the form $f(x+ct) + g(x-ct)$. The zero initial data implies

$$f(x) + g(x) = 0$$
$$cf'(x) - cg'(x) = 0$$

for all $x > 0$. Differentiating the first equation we get

$$f'(x) + g'(x) = 0$$
$$cf'(x) - cg'(x) = 0,$$

from which it follows that $f'(x) = g'(x) = 0$ for $x > 0$. Hence $f(y)$ and $g(y)$ are constant for $y > 0$, and by the initial data $f = -g$. Hence $f(y) = C$ and $g(y) = -C$ for some constant C, for all positive y. Thus, for $x > ct$ the solution is $u(x,t) = C - C = 0$. The boundary data implies $f'(ct) + g'(-ct) = k(t)$. Since $ct > 0$, $f'(ct) = 0$, so $g'(-ct) = k(t)$. Thus $g'(y) = k(-y/c)$ for $y < 0$. By the Fundamental Theorem of Calculus,

$$g(y) - g(0) = \int_0^y k(-s/c)\,ds$$

for $y < 0$, and since $g(0) = -C$,

$$g(y) = -C + \int_0^y k(-s/c)\,ds = -C - c\int_0^{-y/c} k(s)\,ds.$$

Finally, since $x + ct > 0$ for all x and t, $f(x+ct) = C$, so

$$u(x,t) = f(x+ct) + g(x-ct) = C - C - c\int_0^{t-x/c} k(s)\,ds.$$

Thus

$$u(x,t) = \begin{cases} 0 & x > ct \\ -c\displaystyle\int_0^{t-x/c} k(s)\,ds & x < ct. \end{cases}$$

Solution 2. Let $v(x,t) = u(x,t) - xk(t)$. Then v satisfies

$$
\begin{aligned}
v_{tt} - c^2 v_{xx} &= -xk''(t) & & 0 < x < \infty,\ 0 < t < \infty \\
v(x,0) &= -xk(0) & & 0 < x < \infty \\
v_t(x,0) &= -xk'(0) & & 0 < x < \infty \\
v_x(0,t) &= 0 & & 0 < t < \infty.
\end{aligned}
$$

By Duhamel's Principle the solution is

$$v(t) = \mathscr{S}(t)\begin{bmatrix} -xk(0) \\ -xk'(0) \end{bmatrix} + \int_0^t \mathscr{S}(t-s)\begin{bmatrix} 0 \\ -xk''(s) \end{bmatrix}ds,$$

where $\mathscr{S}$ is the solution operator for the homogeneous wave equation with homogeneous Neumann boundary data.

For $x > ct$ the solution operator is given by the usual d'Alembert formula (2.1.8). That is,

$$\left(\mathscr{S}(t)\begin{bmatrix}\phi\\\psi\end{bmatrix}\right)(x) = \frac{1}{2}[\phi(x+ct)+\phi(x-ct)] + \frac{1}{2c}\int_{x-ct}^{x+ct}\psi(y)\,dy.$$

So

$$v(x,t) = \frac{1}{2}[-(x+ct)k(0)-(x-ct)k(0)] + \frac{1}{2c}\int_{x-ct}^{x+ct}-yk'(0)\,dy$$

$$+ \frac{1}{2c}\int_0^t\int_{x-c(t-s)}^{x+c(t-s)}-yk''(s)\,dy\,ds$$

$$= -xk(0) - \frac{y^2k'(0)}{4c}\Big|_{x-ct}^{x+ct} - \frac{1}{4c}\int_0^t y^2k''(s)\Big|_{x-c(t-s)}^{x+c(t-s)}\,ds$$

$$= -xk(0) - xtk'(0) - x\int_0^t(t-s)k''(s)\,ds$$

$$= -xk(0) - xtk'(0) - x[k'(s)(t-s)+k(s)]\Big|_0^t$$

$$= -xk(t).$$

For $x < ct$ the solution operator is given by the formula in the solution to Exercise 3.2.1. That is,

$$\left(\mathscr{S}(t)\begin{bmatrix}\phi\\\psi\end{bmatrix}\right)(x) = \frac{1}{2}[\phi(x+ct)+\phi(ct-x)] + \frac{1}{2c}\left[\int_0^{ct-x}\psi(y)\,dy + \int_0^{ct+x}\psi(y)\,dy\right].$$

Applying this, we get

$$v(x,t) = \frac{1}{2}[-(x+ct)k(0)-(ct-x)k(0)] + \frac{1}{2c}\left[\int_0^{ct-x}-yk'(0)\,dy + \int_0^{ct+x}-yk'(0)\,dy\right]$$

$$+ \frac{1}{2c}\int_0^{t-x/c}\left[\int_0^{c(t-s)-x}-yk''(s)\,dy + \int_0^{c(t-s)+x}-yk''(s)\,dy\right]\,ds$$

$$+ \frac{1}{2c}\int_{t-x/c}^t\int_{x-c(t-s)}^{x+c(t-s)}-yk''(s)\,dy\,ds.$$

The first term simplifies to $A_1 = -ctk(0)$. Next,

$$A_2 = \frac{1}{2c}\left[\int_0^{ct-x}-yk'(0)\,dy + \int_0^{ct+x}-yk'(0)\,dy\right]$$

$$= -\frac{k'(0)}{2c}\left(\frac{1}{2}y^2\Big|_0^{ct-x} + \frac{1}{2}y^2\Big|_0^{ct+x}\right)$$

$$= -\frac{k'(0)}{2c}(x^2+c^2t^2).$$

The third term is

$$A_3 = \frac{1}{2c} \int_0^{t-x/c} \left[\int_0^{c(t-s)-x} -yk''(s)\,dy + \int_0^{c(t-s)+x} -yk''(s)\,dy \right] ds$$

$$= -\frac{1}{2c} \int_0^{t-x/c} k''(s) \left(\frac{1}{2}y^2 \Big|_0^{c(t-s)-x} + \frac{1}{2}y^2 \Big|_0^{c(t-s)+x} \right) ds$$

$$= -\frac{1}{2c} \int_0^{t-x/c} k''(s)[x^2 + c^2(t-s)^2]\,ds.$$

Integrating by parts twice, this becomes

$$A_3 = -\frac{1}{2c} \left(k'(s)[x^2 + c^2(t-s)^2] + 2c^2 k(s)(t-s) \right) \Big|_0^{t-x/c} - \frac{1}{2c} \int_0^{t-x/c} 2c^2 k(s)\,ds$$

$$= -\frac{x^2}{c} k'(t-x/c) - xk(t-x/c) + \frac{k'(0)}{2c}(x^2 + c^2 t^2) + ctk(0) - c \int_0^{t-x/c} k(s)\,ds.$$

The final term is

$$A_4 = \frac{1}{2c} \int_{t-x/c}^t \int_{x-c(t-s)}^{x+c(t-s)} -yk''(s)\,dy\,ds$$

$$= -\frac{1}{4c} \int_{t-x/c}^t k''(s)y^2 \Big|_{x-c(t-s)}^{x+c(t-s)}\,ds$$

$$= -x \int_{t-x/c}^t k''(s)(t-s)\,ds.$$

Integrating by parts gives

$$A_4 = -x \left(k'(s)(t-s) + k(s) \right) \Big|_{t-x/c}^t$$

$$= -xk(t) + \frac{x^2}{c} k'(t-x/c) + xk(t-x/c).$$

Summing A_1 through A_4, we see that

$$v(x,t) = -xk(t) - c \int_0^{t-x/c} k(s)\,ds.$$

Thus

$$u(x,t) = \begin{cases} 0 & x > ct \\ -c \int_0^{t-x/c} k(s)\,ds & x < ct. \end{cases}$$

Section 3.5

3.5.1. We write out the proof for x^+. The same technique works for x^-. Making the change of variables $y = p/2$, we see that

$$\frac{1}{\sqrt{4\pi}} \int_0^\infty e^{-p^2/4} \, dp = \frac{1}{\sqrt{\pi}} \int_0^\infty e^{-y^2} \, dy = \frac{1}{2}.$$

Therefore,

$$\frac{1}{\sqrt{4\pi}} \int_0^\infty e^{-p^2/4} \phi(x^+) \, dp = \frac{1}{2} \phi(x^+).$$

Using this equality, showing that

$$\lim_{t \to 0^+} \left[\frac{1}{\sqrt{4\pi}} \int_0^\infty e^{-p^2/4} \phi(x + \sqrt{k t} p) \, dp \right] = \frac{1}{2} \phi(x^+),$$

is equivalent to showing that

$$\lim_{t \to 0^+} \left[\int_0^\infty e^{-p^2/4} [\phi(x + \sqrt{k t} p) - \phi(x^+)] \, dp \right] = 0.$$

In particular, we need to show that for all $\epsilon > 0$, there exists $\delta > 0$ such that

$$\left| \int_0^\infty e^{-p^2/4} [\phi(x + \sqrt{k t} p) - \phi(x^+)] \, dp \right| < \epsilon$$

if $0 < t < \delta$. Fix $\epsilon > 0$. We will separate the integral into two pieces. First, we consider

$$\int_R^\infty e^{-p^2/4} [\phi(x + \sqrt{k t} p) - \phi(x^+)] \, dp.$$

We note that

$$\left| \int_R^\infty e^{-p^2/4} [\phi(x + \sqrt{k t} p) - \phi(x^+)] \, dp \right| < \frac{\epsilon}{2},$$

as long as R is chosen sufficiently large (and ϕ does not grow too rapidly at infinity). Now for such a choice of R, consider

$$\int_0^R e^{-p^2/4} [\phi(x + \sqrt{k t} p) - \phi(x^+)] \, dp.$$

We note that

$$\left| \int_0^R e^{-p^2/4} [\phi(x + \sqrt{k t} p) - \phi(x^+)] \, dp \right| \leq \max_{0 \leq p \leq R} |\phi(x + \sqrt{k t} p) - \phi(x^+)| \int_0^R e^{-p^2/4} \, dp$$

$$\leq C \max_{0 \leq p \leq R} |\phi(x + \sqrt{k t} p) - \phi(x^+)|.$$

By assumption ϕ is piecewise continuous. Therefore, for all $\tilde{\epsilon} > 0$ there exists a $\tilde{\delta} > 0$ such that

$$|\phi(x^+) - \phi(y)| < \tilde{\epsilon}$$

30

if $0 < y - x^+ < \tilde{\delta}$. Choose $\tilde{\delta} > 0$ such that

$$|\phi(y) - \phi(x^+)| < \frac{\epsilon}{2C},$$

for $0 < y - x < \tilde{\delta}$. Let $\delta = \tilde{\delta}^2/kR^2$. For $0 < t < \delta$, we have $\sqrt{kt}R < \tilde{\delta}$, which implies

$$0 < x + \sqrt{kt}p - x < \tilde{\delta}$$

for all $p \in [0, R]$. And, therefore,

$$\max_{0 \leq p \leq R} |\phi(x + \sqrt{kt}p) - \phi(x^+)| < \frac{\epsilon}{2C}.$$

We conclude that for R chosen sufficiently large above,

$$\left| \int_0^\infty e^{-p^2/4} [\phi(x + \sqrt{kt}p) - \phi(x^+)] \, dp \right| \leq \left| \int_0^R e^{-p^2/4} [\phi(x + \sqrt{kt}p) - \phi(x^+)] \, dp \right|$$

$$+ \left| \int_R^\infty e^{-p^2/4} [\phi(x + \sqrt{kt}p) - \phi(x^+)] \, dp \right|$$

$$< C \frac{\epsilon}{2C} + \frac{\epsilon}{2} = \epsilon,$$

as long as $0 < t < \delta$.

Chapter 4

Section 4.1

4.1.2. We need to solve

$$
\begin{cases}
u_t - k u_{xx} = 0 & 0 < x < l, \ t > 0 \\
u(x,0) = 1 & 0 < x < l \\
u(0,t) = 0 = u(l,t) & t > 0.
\end{cases}
$$

Looking for a solution of the form $u(x,t) = X(x)T(t)$ implies

$$
\frac{T'}{kT} - \frac{X''}{X} = 0 \implies \frac{T'}{kT} = \frac{X''}{X} = -\lambda
$$

where λ is a constant. So we consider the eigenvalue problem

$$
\begin{cases}
X'' = -\lambda X & 0 < x < l \\
X(0) = 0 = X(l).
\end{cases}
$$

If $\lambda = \beta^2 > 0$, then $X(x) = C\cos(\beta x) + D\sin(\beta x)$. The boundary conditions $X(0) = 0 = X(l)$ imply that $C = 0$ and $\beta_n = (n\pi/l)^2$ for $n = 1, 2, \dots$. All eigenvalues are positive (see text). Solving

$$
\frac{T_n'}{kT_n} = -\left(\frac{n\pi}{l}\right)^2,
$$

we see that

$$
T_n(t) = A_n e^{-k\left(\frac{n\pi}{l}\right)^2 t}.
$$

Therefore,

$$
u(x,t) = \sum_{n=1}^{\infty} A_n \sin\left(\frac{n\pi}{l}x\right) e^{-k\left(\frac{n\pi}{l}\right)^2 t}.
$$

The initial condition

$$
u(x,0) = 1 = \frac{4}{\pi}\left(\sin\frac{\pi x}{l} + \frac{1}{3}\sin\frac{3\pi x}{l} + \frac{1}{5}\sin\frac{5\pi x}{l} + \dots\right)
$$

implies

$$
A_n = \begin{cases} \frac{4}{n\pi} & n \text{ odd} \\ 0 & n \text{ even.} \end{cases}
$$

Therefore,

$$
u(x,t) = \frac{4}{\pi} \sum_{n \text{ odd}} \frac{1}{n} \sin\left(\frac{n\pi}{l}x\right) e^{-k\left(\frac{n\pi}{l}\right)^2 t}.
$$

4.1.3. Using separation of variables, we look for a solution of the form $u(x,t) = X(x)T(t)$. Plugging a function of this form into our PDE, we see that X and T must satisfy

$$
\frac{T'}{iT} = \frac{X''}{X} = -\lambda,
$$

32

which implies that X must satisfy the eigenvalue problem

$$\begin{cases} X'' = -\lambda X & 0 < x < l \\ X(0) = 0 = X(l). \end{cases}$$

The eigenfunctions are $X_n(x) = \sin\left(\frac{n\pi}{l}x\right)$, with corresponding eigenvalues $\lambda_n = \left(\frac{n\pi}{l}\right)^2$. Now solving our equation for T, we have

$$T_n' = -i\left(\frac{n\pi}{l}\right)^2 T_n$$

which implies

$$T_n(t) = A_n e^{-i\left(\frac{n\pi}{l}\right)^2 t}.$$

Therefore,

$$u(x,t) = \sum_{n=1}^{\infty} A_n \sin\left(\frac{n\pi}{l}x\right) e^{-i\left(\frac{n\pi}{l}\right)^2 t}.$$

Section 4.2

4.2.1. Separation of variables leads to the eigenvalue problem

$$\begin{cases} X'' = -\lambda X & 0 < x < l \\ X(0) = 0 = X'(l). \end{cases}$$

Looking for positive eigenvalues $\lambda = \beta^2 > 0$ implies

$$X(x) = C\cos(\beta x) + D\sin(\beta x).$$

The boundary condition $X(0) = 0$ implies $C = 0$, while the boundary condition $X'(l) = 0$ implies $\beta = \left(n + \frac{1}{2}\right)\pi/l$. Therefore,

$$\lambda_n = \left(\frac{\left(n + \frac{1}{2}\right)\pi}{l}\right)^2 \qquad X_n(x) = \sin\left(\frac{\left(n + \frac{1}{2}\right)\pi x}{l}\right).$$

If $\lambda = -\gamma^2 < 0$, then $X(x) = A\cosh(\gamma x) + B\sinh(\gamma x)$, and the boundary condition $X(0) = 0$ implies $A = 0$, so $X(x) = B\sinh(\gamma x)$. But the boundary condition $X'(l) = 0$ implies $B\cosh(\gamma l) = 0$, so $B = 0$. Therefore there are no negative eigenvalues.

If $\lambda = 0$, then $X(x) = Ax + B$, and the boundary condition $X(0) = 0$ implies $B = 0$. But then the boundary condition $X'(l) = 0$ implies $A = 0$. Therefore zero is not an eigenvalue, so all the eigenvalues are positive.

Solving the equation $T_n' = -\lambda_n k T_n$ gives $T_n(t) = C_n e^{-k\lambda_n t}$. Therefore,

$$u(x,t) = \sum_{n=0}^{\infty} C_n \sin\left(\frac{\left(n + \frac{1}{2}\right)\pi x}{x}\right) e^{-k\left(n + \frac{1}{2}\right)^2 \pi^2 t/l^2}.$$

4.2.4

(a) Separation of variables leads to the eigenvalue problem

$$\begin{cases} X'' = -\lambda X & 0 < x < l \\ X(-l) = X(l) \\ X'(-l) = X'(l). \end{cases}$$

Looking for positive eigenvalues $\lambda = \beta^2 > 0$ implies

$$X(x) = C\cos(\beta x) + D\sin(\beta x).$$

The boundary condition $X(-l) = X(l)$ implies $D\sin(\beta l) = 0$, so $\beta = \frac{n\pi}{l}$. The boundary condition $X'(-l) = X'(l)$ implies $C\beta\sin(\beta l) = 0$, so $\beta = \frac{n\pi}{l}$. Therefore,

$$\lambda_n = \left(\frac{n\pi}{l}\right)^2 \qquad X_n(x) = C_n\cos\left(\frac{n\pi x}{l}\right) + D_n\sin\left(\frac{n\pi x}{l}\right)$$

for $n = 1, 2, 3, \ldots$

If $\lambda = 0$, then $X(x) = C + Dx$. The boundary condition $X(-l) = X(l)$ implies $D = 0$. The boundary condition $X'(-l) = X'(l) = 0$ will be satisfied for arbitrary C. Therefore, $\lambda = 0$ is an eigenvalue with corresponding eigenfunction $X(x) = C$.

If $\lambda = -\gamma^2 < 0$, then $X(x) = C\cosh(\gamma x) + D\sinh(\gamma x)$. The boundary condition $X(-l) = X(l)$ implies $D\sinh(\gamma l) = 0$, so $D = 0$. The boundary condition $X'(-l) = X'(l)$ implies $C\gamma\sinh(\gamma l) = 0$, so $C = 0$. Therefore, all the eigenvalues are $\lambda_n = (n\pi/l)^2$ for $n = 0, 1, 2, 3, \ldots$

(b) Solving the equation $T_n' = -\lambda_n k T_n$ gives $T_n(t) = A_n e^{-k\lambda_n t}$. Therefore,

$$u(x, t) = \frac{1}{2}A_0 + \sum_{n=1}^{\infty}\left(A_n\cos\left(\frac{n\pi x}{l}\right) + B_n\sin\left(\frac{n\pi x}{l}\right)\right)e^{-kn^2\pi^2 t/l^2}.$$

Section 4.3

4.3.1. First consider positive eigenvalues $\lambda = \beta^2$. The corresponding eigenfunctions are $X(x) = C\cos(\beta x) + D\sin(\beta x)$. The condition $X(0) = 0$ implies $C = 0$, so $X(x) = D\sin(\beta x)$. The condition $X'(l) + aX(l) = 0$ implies $D\beta\cos(\beta l) + aD\sin(\beta l) = 0$. Thus $\tan(\beta l) = -\frac{\beta}{a}$. The eigenvalues are determined by the points of intersection of the graphs of $y = \tan(\beta l)$ and $y = -\beta/a$. See Figure 7.

If $\lambda = 0$, then $X(x) = C + Dx$. The boundary condition $X(0) = 0$ implies $C = 0$. Then $X'(l) + aX(l) = D + aDl = 0$ implies $D(1 + al) = 0$. If $a = -1/l$, then $\lambda = 0$ is an eigenvalue with corresponding eigenfunction $X(x) = Dx$. If $a \neq -1/l$, then $\lambda = 0$ is not an eigenvalue.

Next consider negative eigenvalues $\lambda = -\gamma^2$. The corresponding eigenfunctions are $X(x) = C\cosh\gamma x + D\sinh\gamma x$. The condition $X(0) = 0$ again implies $C = 0$, so $X(x) =$

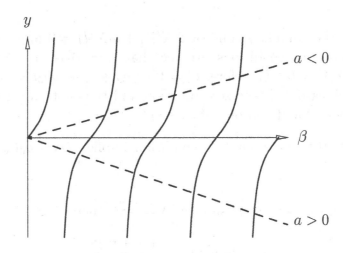

Figure 7: Graphs of $y = \tan(\beta l)$ and $y = -\beta/a$.

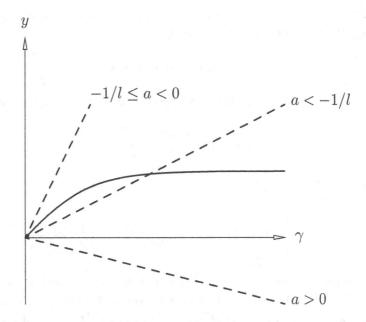

Figure 8: Graphs of $y = \tanh(\gamma l)$ and $y = -\frac{\gamma}{a}$.

$D \sinh \gamma x$. The condition $X'(l) + aX(l) = 0$ implies $D\gamma \cosh(\gamma l) + aD \sinh(\gamma l) = 0$. Thus $\tanh(\gamma l) = -\frac{\gamma}{a}$. First we consider the case when $a > 0$. In this case the left side is positive for $\gamma > 0$ and the right side is negative for $\gamma > 0$. Therefore, if $a > 0$ there are no negative eigenvalues. Second we consider the case when $a < 0$. In this case we will have one negative eigenvalue if $(\tanh(\gamma l))' > (-\gamma/a)'$ at $\gamma = 0$. Since $(\tanh(\gamma l))' = l\,\mathrm{sech}^2(\gamma l)$ and $(-\gamma/a)' = -1/a$, we will have one negative eigenvalue when $l > -1/a$. That is, we will have one negative eigenvalue when $a < -1/l$ with corresponding eigenvector $X(x) = \sinh(\gamma x)$. See Figure 8.

4.3.2.

(a) If $\lambda = 0$, then $X(x) = C + Dx$. The boundary condition $X'(0) - a_0 X(0) = 0$ implies

35

$D = a_0 C$, and the boundary condition $X'(l) + a_l X(l) = 0$ implies $D + a_l[C + Dl] = 0$. Combining these two conditions, we see that $\lambda = 0$ is an eigenvalue if and only if $C[a_0 + a_l + a_l a_0 l] = 0$. Therefore, $C = 0$ or $a_0 + a_l + a_l a_0 l = 0$. But, $C = 0$ implies $D = 0$ in which case X is the zero function and thus not an eigenfunction. Therefore, $\lambda = 0$ is an eigenvalue if and only if $a_0 + a_l = -a_0 a_l l$.

(b) From part (a), the first boundary condition implies $D = a_0 C$. Therefore, $X_0(x) = C(a_0 x + 1)$.

4.3.4 As derived in the text, if we suppose $\lambda = -\gamma^2$, then

$$\tanh \gamma l = -\frac{(a_0 + a_l)\gamma}{\gamma^2 + a_0 a_l}.$$

Let $f(\gamma) = \tanh \gamma l$ and $g(\gamma) = -(a_0 + a_l)\gamma/(\gamma^2 + a_0 a_l)$. Then $f'(0) = l$ and $g'(0) = -(a_0 + a_l)/a_0 a_l$. By the hypothesis, we have

$$-(a_0 + a_l)/a_0 a_l < l.$$

So for γ in a small interval of the form $(0, \delta)$ we have $f(\gamma) > g(\gamma)$. Next notice that

$$g'(\gamma) = -\frac{(a_0 a_l - \gamma^2)(a_0 + a_l)}{(\gamma^2 + a_0 a_l)^2}$$

vanishes at $\gamma = \sqrt{a_0 a_l}$, and

$$g(\sqrt{a_0 a_l}) = -\frac{(a_0 + a_l)\sqrt{a_0 a_l}}{2 a_0 a_l} = \frac{|a_0| + |a_l|}{2\sqrt{|a_0||a_l|}}.$$

Applying the inequality $2ab \leq a^2 + b^2$ with $a = \sqrt{|a_0|}$ and $b = \sqrt{|a_l|}$, it follows that $g(\sqrt{a_0 a_l}) \geq 1$. Thus the maximum of $g(\gamma)$ lies above the line $y = 1$. Since $\tanh x < 1$ for all x, we therefore have $g(\sqrt{a_0 a_l}) > f(\sqrt{a_0 a_l})$. Therefore, there exists some γ_1 between 0 and $\sqrt{a_0 a_l}$ such that $f(\gamma_1) = g(\gamma_1)$. Now since

$$\lim_{\gamma \to \infty} f(\gamma) = 1 \qquad \text{and} \qquad \lim_{\gamma \to \infty} g(\gamma) = 0$$

it follows that $f(\gamma) > g(\gamma)$ for sufficiently large γ. Thus, there is another crossing point, that is, some $\gamma_2 > \sqrt{a_0 a_l}$ such that $f(\gamma_2) = g(\gamma_2)$. Thus both $-\gamma_1^2$ and $-\gamma_2^2$ are negative eigenvalues.

4.3.8

(a) The equation $a_0 + a_l = -l a_0 a_l$ for the hyperbola can also be written as

$$(l a_0 + 1)(l a_l + 1) = 1.$$

This is like the standard hyperbola $a_0 a_l = 1$ but with the origin shifted to the point $(-\frac{1}{l}, -\frac{1}{l})$. See Figure 9.

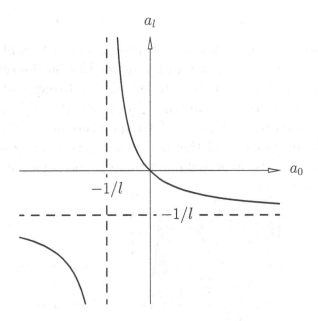

Figure 9: The hyperbola $a_0 + a_l = -a_0 a_l l$.

(b) In order to determine any negative eigenvalues, use the same method as in the text. That is, look for the intersection of the graphs of $y = \tanh \gamma l$ and

$$y = f(\gamma) \equiv -\frac{(a_0 + a_l)\gamma}{\gamma^2 + a_0 a_l}$$

for $\gamma > 0$. The picture of $y = f(x)$ depends on the signs of $a_0 + a_l$ and $a_0 a_l$.

In the first quadrant $a_0 > 0, a_l > 0$, $y = f(\gamma) < 0$, so its graph lies entirely below the γ-axis. So it cannot intersect the tanh curve and there is no negative eigenvalue.

Now consider the wedge-shaped region $a_0 + a_l > 0$ with a_0 and a_l of opposite signs. Then the graph of $y = f(\gamma)$ rises to a vertical asymptote at $\gamma = \sqrt{-a_0 a_l}$ and then reappears below the γ-axis. In case $a_0 + a_l < -la_0 a_l$, this is exactly the graph labeled (18) in Figure 4.3.4, in which case there is exactly one negative eigenvalue. In case $a_0 + a_l > -la_0 a_l$, check that the slope of $f(\gamma)$ is larger than the slope of $y = \tanh l\gamma$ and therefore there is no intersection and no eigenvalue. In case $a_0 + a_l = -la_0 a_l$, there is no negative eigenvalue but there is a zero eigenvalue (as in Exercise 4.3.2).

Now consider the wedge-shaped region $a_0 + a_l < 0$ with a_0 and a_l of opposite signs. The graph of $y = f(\gamma)$ descends to a vertical asymptote at $\gamma = \sqrt{-a_0 a_l}$ and then reappears above the γ-axis. So there is an intersection at a point where $\gamma > \sqrt{-a_0 a_l}$ and there is a single eigenvalue.

The borderline case when $a_0 + a_l = 0$ must be treated slightly differently to avoid dividing by zero. In this case there is the single eigenvalue $\gamma = \sqrt{-a_0 a_l}$.

Finally consider the case when both a_0 and a_l are negative. Then the graph of $y = f(\gamma)$ starts at the origin, rises to a single maximum and descends asymptotically to zero. It is convenient to solve the equation $f(\gamma) = 1$, which can be rewritten using a bit of algebra as $(\gamma + a_0)(\gamma + a_l) = 0$. Thus the graph of $y = f(\gamma)$ crosses the horizontal line $y = 1$

37

at $\gamma = -a_0$ and at $\gamma = -a_l$. Therefore it also crosses the graph of $y = \tanh l\gamma$ once or twice. Now there are only two possibilities, depending on the relative sizes of the slopes $f'(0) = -\frac{a_0 + a_l}{a_0 a_l}$ and $(\tanh l\gamma)'(0) = l$. If $f'(0)$ is the bigger one, then there is only one intersection (at the descent) and only one eigenvalue. If $f'(0)$ is the smaller one, then we are exactly in the situation of Exercise 4.3.4 and there are two intersections, one to each side of the maximum point, and therefore exactly two eigenvalues. If they are equal, then we are exactly on the hyperbola and there is one negative eigenvalue and one zero eigenvalue. See Figure 10.

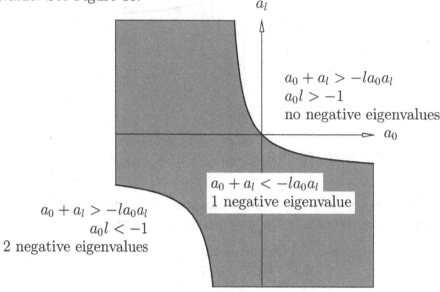

Figure 10: Regions of zero, one or two negative eigenvalues.

(c) The Neumann BC at the end $x = l$ corresponds to the case $a_l = 0$, which is the horizontal axis. Similarly at the other end.

(d) The Dirichlet BC is not a special case of the Robin BC. However, it (at the end $x = l$) corresponds to the limit $a_l \to +\infty$ in the (a_0, a_l)-plane.

4.3.9.

(a) For $\lambda = 0$, $X(x) = C + Dx$. Both boundary conditions imply $C + D = 0$. Therefore, $X(x) = C(1 - x)$, where C is any nonzero constant.

(b) The eigenfunctions are $C \cos \beta x + D \sin \beta x$. The boundary conditions imply $C \cos \beta + D \sin \beta = 0$ and $C + \beta D = 0$, so $D(\sin \beta - \beta \cos \beta) = 0$, or $\tan \beta = \beta$.

(c) The graphs of $y = \tan \beta$ and $y = \beta$ have infinitely many points of intersection. See Figure 11.

(d) Suppose $\lambda = -\gamma^2$. Then

$$X(x) = C \cosh \gamma x + D \sinh \gamma x$$

38

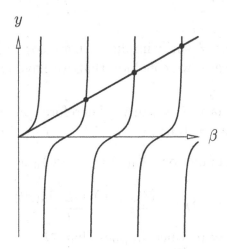

Figure 11: The graphs of $y = \tan\beta$ and $y = \beta$.

and the boundary conditions imply $C\cosh\gamma + D\sinh\gamma = 0$ and $C + D\gamma = 0$, so $D(\sinh\gamma - \gamma\cosh\gamma) = 0$, or $\tanh\gamma = \gamma$. Since $\tanh\gamma < \gamma$ for all positive γ, there are no non-zero solutions of this equation, and thus no negative eigenvalues.

4.3.12.

(a) If $\lambda = 0$, then $v(x) = C + Dx$ which implies $v'(x) = D$. The boundary condition $v'(0) = v'(l) = \frac{v(l)-v(0)}{l}$ therefore implies $D = D = \frac{[C+Dl]-C}{l}$, a condition which is satisfied for all C, D. Therefore, $v(x) = C + Dx$ is an eigenfunction for all constants C, D. In particular, $\lambda = 0$ has two linearly independent eigenfunctions $X_0(x) = 1$ and $Y_0(x) = x$.

(b) If $\lambda = \beta^2 > 0$, then $v(x) = C\cos(\beta x) + D\sin(\beta x)$. The boundary condition $v'(0) = v'(l) = \frac{v(l)-v(0)}{l}$ implies

$$D\beta = -C\beta\sin(\beta l) + D\beta\cos(\beta l) = \frac{[C\cos(\beta l) + D\sin(\beta l)] - C}{l}.$$

The first equality implies that $D(1 - \cos(\beta l)) = -C\sin(\beta l)$. The second equality implies that $D(\beta l - \sin(\beta l)) = C(\cos(\beta l) - 1)$. So to find a solution $(C, D) \neq (0,0)$ requires the equality

$$(1 - \cos(\beta l))^2 = (\beta l - \sin(\beta l))\sin(\beta l).$$

Multiplying this out, it can be written as

$$2(1 - \cos(\beta l)) = \beta l \sin(\beta l).$$

(c) Let $\gamma = \beta l/2$. Therefore,

$$\beta\sin(\beta l) = \frac{2 - 2\cos(\beta l)}{l} \implies \frac{\beta l}{2}\sin(\beta l) = 1 - \cos(\beta l)$$
$$\implies \gamma\sin(2\gamma) = 1 - \cos(2\gamma)$$
$$\implies 2\gamma\sin\gamma\cos\gamma = 1 - [1 - 2\sin^2\gamma]$$
$$\implies \gamma\sin\gamma\cos\gamma = \sin^2\gamma.$$

39

(d) The equation $\gamma \sin\gamma\cos\gamma = \sin^2\gamma$ implies either $\gamma\cos\gamma = \sin\gamma$ or $\sin\gamma = 0$. If $\sin\gamma = 0$, then $\gamma = n\pi$. In this case, we conclude that we have eigenvalues $\lambda = \beta^2 = (2\gamma/l)^2 = (2n\pi/l)^2$.

In the other case, we have $\gamma = \tan\gamma$. Therefore, our other eigenvalues are given by $\lambda = \beta^2 = (2\gamma/l)^2$ where $\gamma = \tan\gamma$.

(e) For $\lambda = 0$, we have eigenfunctions $X_0(x) = 1$ and $Y_0(x) = x$. If $\lambda = \beta^2 = (2n\pi/l)^2$, then using the equality

$$D\beta = \frac{C\cos(\beta l) + D\sin(\beta l) - C}{l},$$

we see that $D(2n\pi/l) = 0$ which implies that $D = 0$. Therefore, the eigenfunctions corresponding to $\lambda_n = (2n\pi/l)^2$ are given by

$$X_n(x) = \cos\left(\frac{2n\pi}{l}x\right).$$

Our other eigenvalues are given by $\lambda = \beta^2 = (2\gamma/l)^2$ where $\gamma = \tan(\gamma)$. The condition $D(1 - \cos(\beta l)) = -C\sin(\beta l)$ implies the corresponding eigenfunctions are given by

$$Y_n(x) = \cos(\beta x) - \left(\frac{\sin(\beta l)}{1 - \cos(\beta l)}\right)\sin(\beta x) = \cos(\beta x) - \frac{2}{\beta l}\sin(\beta x)$$

by part (b).

(f) Now to solve the heat equation with the boundary conditions specified above, we use separation of variables to lead us to the eigenvalue problem stated above. To recap, our eigenvalues and eigenfunctions are given by

$$\lambda_0 = 0 \text{ with } X_0(x) = 1, Y_0(x) = x$$

$$\lambda_n = \left(\frac{2n\pi}{l}\right)^2 \text{ with } X_n(x) = \cos\left(\frac{2n\pi}{l}x\right)$$

$$\alpha_n = \beta_n^2 \text{ where } \beta_n = \frac{2\gamma_n}{l} \text{ for } \gamma_n = \tan(\gamma_n)$$

$$\text{with } Y_n(x) = \cos(\beta_n x) - \left(\frac{\sin(\beta_n l)}{1 - \cos(\beta_n l)}\right)\sin(\beta_n x).$$

Our equation for T,

$$T_n' = -k\lambda_n T_n,$$

has solutions

$$T_n(t) = A_n e^{-k\lambda_n t}.$$

Therefore, the solution is

$$u(x,t) = A_0 + B_0 x + \sum_{n=1}^{\infty} C_n \cos\left(\frac{2n\pi}{l}x\right) e^{-k\left(\frac{2n\pi}{l}\right)^2 t}$$

$$+ \sum_{n=1}^{\infty} D_n \left[\cos(\beta_n x) - \frac{2}{\beta_n l}\sin(\beta_n x)\right] e^{-k\beta_n^2 t}$$

where the coefficients A_0, B_0, C_n, D_n are chosen such that $u(x,0) = \phi(x)$.

(g) From our solution in (f), we see that as $t \to +\infty$, all terms decay to zero, except for the term $A_0 + B_0 x$. Therefore,

$$\lim_{t \to +\infty} u(x,t) = A_0 + B_0 x.$$

4.3.15. Our eigenvalue problem can be rewritten as

$$\begin{cases} \kappa_1^2 X'' + \lambda \rho_1^2 X = 0 & 0 < x < a \\ \kappa_2^2 X'' + \lambda \rho_2^2 X = 0 & a < x < l \\ X(0) = 0 = X(l). \end{cases}$$

If $\lambda = \beta^2 > 0$, the solution of our equations is given by

$$X(x) = \begin{cases} A \cos\left(\frac{\beta \rho_1 x}{\kappa_1}\right) + B \sin\left(\frac{\beta \rho_1 x}{\kappa_1}\right) & 0 < x < a \\ C \cos\left(\frac{\beta \rho_2 x}{\kappa_2}\right) + D \sin\left(\frac{\beta \rho_2 x}{\kappa_2}\right) & a < x < l. \end{cases}$$

The boundary condition $X(0) = 0$ implies $A = 0$. The boundary condition $X(l) = 0$ implies $C \cos\left(\frac{\beta \rho_2 l}{\kappa_2}\right) + D \sin\left(\frac{\beta \rho_2 l}{\kappa_2}\right) = 0$. In order to guarantee that our eigenfunction is continuous at $x = a$, we require

$$B \sin\left(\frac{\beta \rho_1 a}{\kappa_1}\right) = C \cos\left(\frac{\beta \rho_2 a}{\kappa_2}\right) + D \sin\left(\frac{\beta \rho_2 a}{\kappa_2}\right).$$

Additionally, to guarantee our eigenfunction has a continuous derivative at $x = a$, we require

$$B \frac{\beta \rho_1}{\kappa_1} \cos\left(\frac{\beta \rho_1 a}{\kappa_1}\right) = -C \frac{\beta \rho_2}{\kappa_2} \sin\left(\frac{\beta \rho_2 a}{\kappa_2}\right) + D \frac{\beta \rho_2}{\kappa_2} \cos\left(\frac{\beta \rho_2 a}{\kappa_2}\right).$$

Eliminating B from the last two equations, we see that

$$\frac{\rho_1}{\kappa_1} \cot\left(\frac{\beta \rho_1 a}{\kappa_1}\right) \left[C \cos\left(\frac{\beta \rho_2 a}{\kappa_2}\right) + D \sin\left(\frac{\beta \rho_2 a}{\kappa_2}\right)\right] = -C \frac{\rho_2}{\kappa_2} \sin\left(\frac{\beta \rho_2 a}{\kappa_2}\right) + D \frac{\rho_2}{\kappa_2} \cos\left(\frac{\beta \rho_2 a}{\kappa_2}\right).$$

Now multiplying this equation by $\cos\left(\frac{\beta \rho_2 l}{\kappa_2}\right)$ and using the relation

$$C \cos\left(\frac{\beta \rho_2 l}{\kappa_2}\right) + D \sin\left(\frac{\beta \rho_2 l}{\kappa_2}\right) = 0,$$

we see that

$$\frac{\rho_1}{\kappa_1} \cot\left(\frac{\beta \rho_1 a}{\kappa_1}\right) \left[-D \sin\left(\frac{\beta \rho_2 l}{\kappa_2}\right) \cos\left(\frac{\beta \rho_2 a}{\kappa_2}\right) + D \sin\left(\frac{\beta \rho_2 a}{\kappa_2}\right) \cos\left(\frac{\beta \rho_2 l}{\kappa_2}\right)\right]$$
$$= \left[D \frac{\rho_2}{\kappa_2} \sin\left(\frac{\beta \rho_2 l}{\kappa_2}\right) \sin\left(\frac{\beta \rho_2 a}{\kappa_2}\right) + D \frac{\rho_2}{\kappa_2} \cos\left(\frac{\beta \rho_2 a}{\kappa_2}\right) \cos\left(\frac{\beta \rho_2 l}{\kappa_2}\right)\right].$$

Using trigonometric identities, we can rewrite this equation as

$$D\frac{\rho_1}{\kappa_1}\cot\left(\frac{\beta\rho_1 a}{\kappa_1}\right)\sin\left(\frac{\beta\rho_2(a-l)}{\kappa_2}\right) = D\frac{\rho_2}{\kappa_2}\cos\left(\frac{\beta\rho_2(a-l)}{\kappa_2}\right).$$

As we do not want $D = 0$, we conclude that in order that $\lambda = \beta^2 > 0$ be an eigenvalue, we must have

$$\frac{\rho_1}{\kappa_1}\cot\left(\frac{\beta\rho_1 a}{\kappa_1}\right) + \frac{\rho_2}{\kappa_2}\cot\left(\frac{\beta\rho_2(l-a)}{\kappa_2}\right) = 0.$$

4.3.16. If $\lambda = \beta^4 > 0$, then $X'''' - \beta^4 X = 0$. The general solution of this ODE is $X(x) = A\cosh(\beta x) + B\sinh(\beta x) + C\cos(\beta x) + D\sin(\beta x)$. The boundary condition $X(0) = 0$ implies $A + C = 0$, and the boundary condition $X(l) = 0$ implies $A\cosh(\beta l) + B\sinh(\beta l) + C\cos(\beta l) + D\sin(\beta l) = 0$. The boundary condition $X''(0) = 0$ implies $A - C = 0$, and the boundary condition $X''(l) = 0$ implies $A\cosh(\beta l) + B\sinh(\beta l) - C\cos(\beta l) - D\sin(\beta l) = 0$. Because $A + C = 0 = A - C$, we have $A = C = 0$. Substituting $A = 0 = C$ into the second and fourth equations above, we have

$$B\sinh(\beta l) + D\sin(\beta l) = 0$$
$$B\sinh(\beta l) - D\sin(\beta l) = 0.$$

Adding these two equations, we have $B\sinh(\beta l) = 0$ which implies $B = 0$ or $\sinh(\beta l) = 0$. Since $\sinh(\beta l) \neq 0$ (because $\beta l \neq 0$), we must have $B = 0$. Therefore, $D\sin(\beta l) = 0$. Since we do not want $D = 0$ (as this would imply $X(x) \equiv 0$), we must have $\sin(\beta l) = 0$. In particular, $\beta = n\pi/l$.

We conclude that our eigenvalues and corresponding eigenfunctions are given by

$$\lambda_n = \left(\frac{n\pi}{l}\right)^4 \qquad X_n(x) = \sin\left(\frac{n\pi}{l}x\right) \qquad n = 1, 2, \ldots.$$

4.3.18

(a) Let $u(x, t) = X(x)T(t)$. Then $XT'' + c^2 X''''T = 0$, so

$$-\frac{T''}{c^2 T} = \frac{X''''}{X} = \text{constant} = \lambda$$

and thus $X'''' = \lambda X$. The boundary conditions are $X(0) = X'(0) = 0$ and $X''(l) = X'''(l) = 0$.

(b) If $\lambda = 0$ then $X(x) = A + Bx + Cx^2 + Dx^3$. The boundary conditions imply $X(0) = X'(0) = 0$ and $X''(l) = X'''(l) = 0$. The first two conditions imply $A = B = 0$. The other conditions imply $2C + 6Dl = 6D = 0$, so $C = D = 0$. So $X(x) \equiv 0$, and zero is not an eigenvalue.

(c) As in Exercise 4.3.16, the eigenfunctions take the form $X(x) = A\cos\beta x + B\sin\beta x + C\cosh\beta x + D\sinh\beta x$. The boundary conditions $X(0) = X'(0) = 0$ imply $A + C = B + D = 0$, so

$$X(x) = A(\cos\beta x - \cosh\beta x) + B(\sin\beta x - \sinh\beta x).$$

42

Thus the conditions $X''(l) = X'''(l) = 0$ imply

$$
\begin{aligned}
A(\cos \beta l + \cosh \beta l) \;+\; B(\sin \beta l + \sinh \beta l) &= 0 \\
A(\sin \beta l - \sinh \beta l) \;+\; B(-\cos \beta l - \cosh \beta l) &= 0.
\end{aligned}
$$

For this system to have a nontrivial solution for A and B, the determinant

$$
\begin{vmatrix}
\cos \beta l + \cosh \beta l & \sin \beta l + \sinh \beta l \\
\sin \beta l - \sinh \beta l & -\cos \beta l - \cosh \beta l
\end{vmatrix}
$$

must vanish. Thus,

$$
-\left[\cos^2 \beta l + \sin^2 \beta l + 2\cos \beta l \cosh \beta l + \cosh^2 \beta l - \sinh^2 \beta l\right] = 0.
$$

Using the identities $\cos^2 x + \sin^2 x = 1 = \cosh^2 x - \sinh^2 x$, we find the eigenvalue equation

$$
\cos \beta l \cosh \beta l = -1.
$$

Chapter 5

Section 5.1

5.1.1. When $x = \pi/4$, the terms $\sin nx$ for n odd repeat the pattern $\sqrt{2}/2$, $\sqrt{2}/2$, $-\sqrt{2}/2$, $-\sqrt{2}/2$. Thus the series takes the form

$$1 = \sum_{n \text{ odd}} \frac{4}{n\pi} \sin(n\pi/4) = \frac{4}{\pi} \frac{\sqrt{2}}{2} \left(1 + \frac{1}{3} - \frac{1}{5} - \frac{1}{7} + \cdots \right).$$

Therefore

$$1 + \frac{1}{3} - \frac{1}{5} - \frac{1}{7} + \cdots = \frac{\pi\sqrt{2}}{4}.$$

5.1.2.

(a) The coefficients of the sine series for ϕ are

$$A_n = 2 \int_0^1 x^2 \sin(n\pi x)\, dx$$

$$= -\frac{2}{n\pi} \cos(n\pi x) x^2 \Big|_0^1 + \frac{4}{n\pi} \int_0^1 x \cos(n\pi x)\, dx$$

$$= -\frac{2}{n\pi} \cos(n\pi) + \frac{4}{n^2\pi^2} \sin(n\pi x) x \Big|_0^1 - \frac{4}{n^2\pi^2} \int_0^1 \sin(n\pi x)\, dx$$

$$= -\frac{2}{n\pi} \cos(n\pi) + \frac{4}{n^3\pi^3} [\cos(n\pi) - 1]$$

$$= -\frac{2}{n\pi} (-1)^n + \frac{4}{n^3\pi^3} [(-1)^n - 1],$$

so

$$x^2 = \sum_{n=1}^{\infty} \left(-\frac{2}{n\pi} (-1)^n + \frac{4}{n^3\pi^3} [(-1)^n - 1] \right) \sin(n\pi x).$$

5.1.4. For $m = 0$,

$$A_0 = \frac{1}{\pi} \int_{-\pi}^{\pi} |\sin x|\, dx = \frac{4}{\pi}.$$

For $m > 0$,

$$A_m = \frac{1}{\pi} \int_{-\pi}^{\pi} |\sin x| \cos mx\, dx = \frac{2}{\pi} \int_0^{\pi} \sin x \cos mx\, dx.$$

Now $\sin a \cos b = \frac{1}{2}(\sin(b+a) - \sin(b-a))$ so the integral becomes

$$\frac{1}{\pi} \int_0^{\pi} \sin(m+1)x - \sin(m-1)x\, dx = \frac{1}{\pi} \left[-\frac{1}{m+1} \cos(m+1)x + \frac{1}{m-1} \cos(m-1)x \right]_0^{\pi}$$

$$= \begin{cases} 0 & m \text{ odd} \\ \frac{-4}{\pi(m^2-1)} & m \text{ even.} \end{cases}$$

44

Thus

$$|\sin x| = \frac{2}{\pi} + \sum_{m \text{ even}} \frac{-4}{\pi(m^2 - 1)} \cos mx$$

Evaluating this at $x = 0$ and letting $m = 2n$ gives

$$\sum_{n=1}^{\infty} \frac{1}{4n^2 - 1} = \frac{1}{2}.$$

Evaluating at $x = \pi/2$ gives

$$1 = \frac{2}{\pi} + \sum_{n=1}^{\infty} \frac{-4(-1)^n}{\pi(4n^2 - 1)},$$

so

$$\sum_{n=1}^{\infty} \frac{(-1)^n}{4n^2 - 1} = \frac{2 - \pi}{4}.$$

5.1.5.

(a) The sine series for $\phi(x) = x$ is

$$x = \sum_{m=1}^{\infty} \frac{(-1)^{m+1} 2l}{m\pi} \sin \frac{m\pi x}{l}$$

by equation (5.1.12). Integrating term-by-term gives

$$\frac{1}{2}x^2 = C + \sum_{m=1}^{\infty} \frac{(-1)^m 2l^2}{m^2 \pi^2} \cos \frac{m\pi x}{l}.$$

The constant C must be

$$C = \frac{1}{2}A_0 = \frac{1}{l} \int_0^l \frac{1}{2}x^2 \, dx = \frac{l^2}{6},$$

so

$$\frac{1}{2}x^2 = \frac{l^2}{6} + \sum_{m=1}^{\infty} \frac{(-1)^m 2l^2}{m^2 \pi^2} \cos \frac{m\pi x}{l}.$$

(b) Evaluating at $x = 0$ yields

$$\sum_{m=1}^{\infty} \frac{(-1)^{m+1}}{m^2} = \frac{\pi^2}{12}.$$

5.1.9. Since $\phi(x) = 0$ and $\psi(x) = \cos^2 x = \frac{1}{2} + \frac{1}{2}\cos 2x$, the coefficients in the cosine series for ϕ are $A_n = 0$ for all n, and the coefficients in the cosine series for ψ are $B_0 = 1$, $B_2 = \frac{1}{4c}$ and $B_n = 0$ for all other n. Hence equation (4.2.7) implies

$$u(x, t) = \frac{1}{2}t + \frac{1}{4c} \sin 2ct \cos 2x.$$

45

Section 5.2

5.2.1.

(a) The function $\sin(ax)$ is odd and periodic with period $2\pi/a$.

(b) The function e^{ax} is neither even, nor odd nor periodic.

5.2.3. Suppose ϕ is an odd function. Then

$$
\int_{-l}^{l} \phi(x)\,dx = \int_{-l}^{0} \phi(x)\,dx + \int_{0}^{l} \phi(x)\,dx
$$
$$
= -\int_{-l}^{0} \phi(-x)\,dx + \int_{0}^{l} \phi(x)\,dx
$$
$$
= \int_{l}^{0} \phi(y)\,dy + \int_{0}^{l} \phi(x)\,dx
$$
$$
= -\int_{0}^{l} \phi(y)\,dy + \int_{0}^{l} \phi(x)\,dx
$$
$$
= 0.
$$

Next suppose ϕ is even. Then

$$
\int_{-l}^{l} \phi(x)\,dx = \int_{-l}^{0} \phi(x)\,dx + \int_{0}^{l} \phi(x)\,dx
$$
$$
= +\int_{-l}^{0} \phi(-x)\,dx + \int_{0}^{l} \phi(x)\,dx
$$
$$
= -\int_{l}^{0} \phi(y)\,dy + \int_{0}^{l} \phi(x)\,dx
$$
$$
= +\int_{0}^{l} \phi(y)\,dy + \int_{0}^{l} \phi(x)\,dx
$$
$$
= 2\int_{0}^{l} \phi(x)\,dx.
$$

5.2.5. Let $\tilde{\phi}$ be the odd extension of ϕ. By Exercise 5.2.4(a), its full Fourier series consists only of sine terms. Thus

$$
\tilde{\phi}(x) = \sum_{n=1}^{\infty} A_n \sin \frac{n\pi x}{l},
$$

where

$$
A_n = \frac{1}{l} \int_{-l}^{l} \tilde{\phi}(x) \sin \frac{n\pi x}{l}\,dx.
$$

Since the integrand is the product of two odd functions, and is therefore even, by equation (5.2.5) it follows that

$$A_n = \frac{2}{l} \int_0^l \tilde{\phi}(x) \sin \frac{n\pi x}{l} \, dx = \frac{2}{l} \int_0^l \phi(x) \sin \frac{n\pi x}{l} \, dx,$$

since $\tilde{\phi}(x) = \phi(x)$ for $x \in (0, l)$. Thus the coefficients are exactly those of the sine series for ϕ. So when $x \in (0, l)$, the series above is exactly the sine series for ϕ.

5.2.9. Note that having period π is a much stricter condition than having period 2π. Using the orthogonality of $\sin nx$ on $(-\pi, \pi)$, it follows that

$$a_n = \frac{1}{\pi} \int_{-\pi}^{\pi} \phi(x) \sin nx \, dx.$$

Since ϕ is of period π, this becomes

$$a_n = \frac{1}{\pi} \int_{-\pi}^{0} \phi(x) \sin nx \, dx + \frac{1}{\pi} \int_0^{\pi} \phi(x) \sin nx \, dx$$
$$= \frac{1}{\pi} \int_0^{\pi} \phi(y) \sin n(y - \pi) \, dx + \frac{1}{\pi} \int_0^{\pi} \phi(x) \sin nx \, dx$$
$$= \frac{1}{\pi} \int_0^{\pi} \phi(y) \sin ny \cos n\pi \, dx + \frac{1}{\pi} \int_0^{\pi} \phi(x) \sin nx \, dx.$$

For n odd, $\cos n\pi = -1$, so

$$a_n = -\frac{1}{\pi} \int_0^{\pi} \phi(y) \sin ny \, dx + \frac{1}{\pi} \int_0^{\pi} \phi(x) \sin nx \, dx = 0.$$

5.2.10.

(a) By definition $\phi_{\text{odd}}(x) = -\phi(-x)$ for $-l < x < 0$, and $\phi_{\text{odd}}(x) = \phi(x)$ for $0 < x < l$. So in order for ϕ_{odd} to be continuous at $x = 0$ we need

$$\lim_{x \to 0^+} \phi(x) = -\lim_{x \to 0^-} \phi(-x).$$

The limit on the right is the same as $-\lim_{x \to 0^+} \phi(x)$, so we must have $\lim_{x \to 0^+} \phi(x) = 0$.

(b) By part (a) we require

$$\lim_{x \to 0^+} \phi(x) = \lim_{x \to l^-} \phi(x) = 0$$

in order for ϕ_{odd} to be continuous. In order for ϕ_{odd} to also be differentiable at $x = 0$ we need

$$\phi'_{\text{odd}}(0) = \lim_{h \to 0} \frac{\phi_{\text{odd}}(h) - \phi_{\text{odd}}(0)}{h} = \lim_{h \to 0} \frac{\phi_{\text{odd}}(h)}{h}$$

to exist. Evaluating this from the right and using L'Hopital's rule, we have

$$\lim_{h \to 0^+} \frac{\phi_{\text{odd}}(h)}{h} = \lim_{h \to 0^+} \frac{\phi(h)}{h} = \lim_{h \to 0^+} \phi'(h).$$

47

On the other hand, the limit from the left is

$$\lim_{k \to 0^-} \frac{\phi_{\text{odd}}(k)}{k} = \lim_{k \to 0^-} \frac{-\phi(-k)}{k} = \lim_{k \to 0^-} \phi'(-k) = \lim_{h \to 0^+} \phi'(h).$$

Since the left and right hand limits agree, $\phi'_{\text{odd}}(0)$ exists provided the limit $\lim_{h \to 0^+} \phi'(h)$ exists.

(The result of part (b) is sometimes summarized informally by saying that ϕ is continuously differentiable on $[0, l]$ and $\phi(0) = 0$.)

5.2.11. The complex form of the Fourier series for e^x is $\sum_{n=-\infty}^{\infty} C_n e^{in\pi x/l}$, where

$$C_n = \frac{1}{2l} \int_{-l}^{l} e^x e^{-in\pi x/l} \, dx = \frac{e^{x(1 - in\pi/l)}}{2l(1 - in\pi/l)} \bigg|_{-l}^{l}$$

$$= \frac{e^{l - in\pi} - e^{-l + in\pi}}{2l(1 - in\pi/l)} = \frac{(-1)^n \sinh(l)}{l(1 - in\pi/l)}$$

$$= \frac{(-1)^n (l + in\pi) \sinh(l)}{l^2 + n^2\pi^2}.$$

The real form is therefore

$$\frac{1}{2}A_0 + \sum_{n=1}^{\infty} A_n \cos(n\pi x/l) + B_n \sin(n\pi x/l)$$

where

$$A_n = C_n + C_{-n} = \frac{2l(-1)^n \sinh(l)}{l^2 + n^2\pi^2}$$

and

$$B_n = i(C_n - C_{-n}) = \frac{(-1)^{n+1} 2n\pi \sinh(l)}{l^2 + n^2\pi^2}.$$

5.2.17. First suppose f is real-valued. Then $\overline{f(x)} = f(x)$ for all x, so

$$\overline{c_n} = \frac{1}{2l} \int_{-l}^{l} \overline{f(x) e^{-in\pi x/l}} \, dx = \frac{1}{2l} \int_{-l}^{l} f(x) \overline{e^{-in\pi x/l}} \, dx$$

$$= \frac{1}{2l} \int_{-l}^{l} f(x) e^{+in\pi x/l} \, dx = \frac{1}{2l} \int_{-l}^{l} f(x) e^{-i(-n)\pi x/l} \, dx = c_{-n}.$$

Conversely, suppose that $c_n = \overline{c_{-n}}$ for every n. Then

$$\overline{f(x)} = \overline{\sum_{n=-\infty}^{\infty} c_n e^{in\pi x/l}} = \sum_{n=-\infty}^{\infty} \overline{c_n}\, \overline{e^{in\pi x/l}}$$

$$= \sum_{n=-\infty}^{\infty} c_{-n}\, e^{-in\pi x/l} = \sum_{m=-\infty}^{\infty} c_m\, e^{im\pi x/l}$$

$$= f(x),$$

so f is real-valued.

Section 5.3

5.3.1.

(a) The set of all vectors orthogonal to $[1,1,1]$ and $[1,-1,0]$ consists of all multiples of the cross product $[1,1,1] \times [1,-1,0] = [1,1,-2]$.

(b) Let $\mathbf{v}_1 = [1,1,1]$, $\mathbf{v}_2 = [1,-1,0]$, $\mathbf{v}_3 = [1,1,-2]$, and let $\mathbf{x} = [2,-3,5]$. Then

$$\mathbf{x} = \frac{\mathbf{x} \cdot \mathbf{v}_1}{\mathbf{v}_1 \cdot \mathbf{v}_1}\mathbf{v}_1 + \frac{\mathbf{x} \cdot \mathbf{v}_2}{\mathbf{v}_2 \cdot \mathbf{v}_2}\mathbf{v}_2 + \frac{\mathbf{x} \cdot \mathbf{v}_3}{\mathbf{v}_3 \cdot \mathbf{v}_3}\mathbf{v}_3 = \frac{4}{3}\mathbf{v}_1 + \frac{5}{2}\mathbf{v}_2 - \frac{11}{6}\mathbf{v}_3.$$

5.3.2.

(b) Let $p(x) = ax^2 + bx + c$. Then

$$(p,1) = \int_{-1}^{1} ax^2 + bx + c\, dx = \frac{2}{3}a + 2c,$$

$$(p,x) = \int_{-1}^{1} ax^3 + bx^2 + cx\, dx = \frac{2}{3}b.$$

Since both of these inner products must vanish, $b = 0$ and $a = -3c$. Choosing $c = -1$ we see that $p(x) = 3x^2 - 1$ is orthogonal to both 1 and x.

5.3.4

(a) Let $v(x,t) = u(x,t) - U$. Then v satisfies the diffusion equation with boundary conditions $v(0,t) = 0$, $v_x(l,t) = 0$ and initial condition $v(x,0) = -U$. The eigenfunctions corresponding to these boundary conditions are $X_n(x) = \sin((n + \frac{1}{2})\pi x/l)$. Since $(X_n, X_n) = l/2$, the coefficients in the expansion of $-U$ in terms of these functions are

$$A_n = \frac{2}{l}\int_0^l -U \sin\left((n + \tfrac{1}{2})\pi x/l\right)\, dx = -\frac{4U}{(2n+1)\pi}.$$

49

Therefore

$$v(x,t) = \sum_{n=0}^{\infty} \frac{-4U}{(2n+1)\pi} \sin\left((n+\tfrac{1}{2})\pi x/l\right) e^{-(n+\frac{1}{2})^2 \pi^2 kt/l^2},$$

so

$$u(x,t) = U - \sum_{n=0}^{\infty} \frac{4U}{(2n+1)\pi} \sin\left((n+\tfrac{1}{2})\pi x/l\right) e^{-(n+\frac{1}{2})^2 \pi^2 kt/l^2}.$$

(c) Since

$$u(l,t) = U - \sum_{n=0}^{\infty} \frac{4U(-1)^n}{(2n+1)\pi} e^{-(n+\frac{1}{2})^2 \pi^2 kt/l^2},$$

and the series is an alternating series of decreasing terms, the difference between $u(l,t)$ and U is bounded by the absolute value of the first term,

$$\frac{4|U|}{\pi} e^{-\pi^2 kt/4l^2}.$$

If we want this to be less than ϵ, then we need

$$t > \frac{4l^2 \ln(4|U|/\pi\epsilon)}{\pi^2 k}.$$

5.3.6. The solution of $X' = \lambda X$ is $X = Ce^{\lambda x}$. The boundary condition $X(0) = X(1)$ therefore implies $e^{\lambda} = 1$, so $\lambda = 2n\pi i$. The eigenfunctions are $X_n = e^{2\pi i n x}$ and since

$$(X_m, X_n) = \int_0^1 X_m(x)\overline{X_n(x)}\, dx = \int_0^1 e^{2\pi i x(m-n)}\, dx = \left[\frac{e^{2\pi i x(m-n)}}{2\pi i(m-n)}\right]_0^1 = 0$$

for $m \neq n$, they are orthogonal.

5.3.10.

(a) We prove this by induction. First, notice that

$$(Y_2, Z_1) = (X_2, Z_1) - (X_2, Z_1)(Z_1, Z_1) = (X_2, Z_1) - (X_2, Z_1) = 0,$$

so Y_2 and Z_1 are orthogonal. Hence Z_2 and Z_1 are orthogonal. Next suppose we know that Z_1 through Z_{k-1} are orthogonal. Vector Y_k is given by

$$Y_k = X_k - \sum_{j=1}^{k-1}(X_k, Z_j)Z_j.$$

So for any $1 \leq i < k$,

$$(Y_k, Z_i) = (X_k, Z_i) - \sum_{j=1}^{k-1}(X_k, Z_j)(Z_j, Z_i) = (X_k, Z_i) - (X_k, Z_i) = 0,$$

where we have used the fact that for $1 \leq i, j < k$, $(Z_j, Z_i) = 0$ if $i \neq j$ and $(Z_j, Z_i) = 1$ if $i = j$. Thus Y_k is orthogonal to Z_i for $1 \leq i < k$. Thus Z_k is also orthogonal to Z_1 through Z_{k-1}. By induction it therefore follows that all the Z_i are mutually orthogonal.

(b) Let $X_1 = \cos x + \cos 2x$ and $X_2 = 3\cos x - 4\cos 2x$. Then

$$\|X_1\|^2 = \int_0^\pi (\cos x + \cos 2x)^2 \, dx$$

$$= \int_0^\pi \cos^2 x + 2\cos x \cos 2x + \cos^2 2x$$

$$= \int_0^\pi 1 + \frac{1}{2}\cos 2x + \frac{1}{2}\cos 4x + \cos 3x + \cos x \, dx$$

$$= \pi,$$

$$Z_1 = \frac{X_1}{\|X_1\|} = \frac{1}{\sqrt{\pi}}(\cos x + \cos 2x).$$

Next,

$$(X_2, Z_1) = \frac{1}{\sqrt{\pi}} \int_0^\pi (3\cos x - 4\cos 2x)(\cos x + \cos 2x) \, dx$$

$$= \frac{1}{\sqrt{\pi}} \int_0^\pi 3\cos^2 x - \cos 2x \cos x - 4\cos^2 2x \, dx$$

$$= \frac{1}{\sqrt{\pi}} \int_0^\pi -\frac{1}{2} + \frac{3}{2}\cos 2x - 2\cos 4x - \frac{1}{2}\cos 3x - \frac{1}{2}\cos x \, dx$$

$$= -\frac{\sqrt{\pi}}{2}$$

so

$$Y_2 = X_2 - (X_2, Z_1)Z_1 = (3\cos x - 4\cos 2x) + \frac{1}{2}(\cos x + \cos 2x) = \frac{7}{2}\cos x - \frac{7}{2}\cos 2x.$$

Finally, by similar calculations to those above, $\|Y_2\| = \frac{7}{2}\sqrt{\pi}$, so

$$Z_2 = \frac{Y_2}{\|Y_2\|} = \frac{1}{\sqrt{\pi}}(\cos x - \cos 2x).$$

5.3.11.

(a) For Dirichlet boundary conditions $f(a) = f(b) = 0$, so $f(x)f'(x)|_a^b = 0$. For Neumann boundary conditions $f'(a) = f'(b) = 0$, so $f(x)f'(x)|_a^b = 0$. For periodic boundary conditions, $f(a) = f(b)$ and $f'(a) = f'(b)$, so that $f(x)f'(x)|_a^b = 0$.

(b) For Robin boundary conditions,

$$f(x)f'(x)|_a^b = f(b)f'(b) - f(a)f'(a) = -a_l f(b)^2 - a_0 f(a)^2 \le 0,$$

provided both a_0 and a_l are positive (or nonnegative).

5.3.12. Let $u = g(x)$, $dv = f''(x)\,dx$. Then $du = g'(x)\,dx$ and $v = f'(x)$, so integrating by parts gives

$$\int_a^b f''(x)g(x)\,dx = \left[g(x)f'(x)\right]_a^b - \int_a^b f'(x)g'(x)\,dx.$$

Section 5.4

5.4.1.

(a) The geometric series has ratio $-x^2$. The Nth partial sum is therefore

$$S_N = \frac{1 - (-x^2)^{N+1}}{1 + x^2}.$$

For each $x \in (-1, 1)$ this converges to $1/(1 + x^2)$, so the series does converge pointwise.

(b) The series does not converge uniformly on $(-1, 1)$. For suppose N is odd. Then it is possible to choose $x_* \in (-1, 1)$ so that $1 - (-x_*^2)^{N+1} < 1/4$, and thus $S_N(x_*) < 1/4$. Since $1/(1 + x^2) > 1/2$ for all $x \in (-1, 1)$, it follows that $|S_N(x_*) - 1/(1 + x_*^2)| > 1/4$, so $\max |S_N(x) - 1/(1 + x^2)| > 1/4$ and thus $\max |S_N(x) - 1/(1 + x^2)| \not\to 0$.

(c) Since

$$\int_{-1}^1 \left| \frac{1 - (-x^2)^{N+1}}{1 + x^2} - \frac{1}{1 + x^2} \right|^2 dx = \int_{-1}^1 \frac{x^{4N+4}}{(1 + x^2)^2}\,dx$$

$$\leq \int_{-1}^1 x^{4N+4}\,dx$$

$$= \frac{2}{4N + 5} \to 0$$

as $N \to \infty$, the series does converge in the L^2 sense.

5.4.6. For $n > 1$,

$$
\begin{aligned}
A_n &= \frac{2}{\pi} \int_0^\pi \cos x \sin(nx)\,dx \\
&= \frac{1}{\pi} \int_0^\pi \sin((n+1)x) + \sin((n-1)x)\,dx \\
&= \left(\frac{-\cos((n+1)x)}{\pi(n+1)} + \frac{-\cos((n-1)x)}{\pi(n-1)} \right) \Bigg|_0^\pi \\
&= \left(\frac{1 + (-1)^n}{\pi(n+1)} + \frac{1 + (-1)^n}{\pi(n-1)} \right) \\
&= \frac{(2n)(1 + (-1)^n)}{\pi(n^2 - 1)}.
\end{aligned}
$$

Since $A_1 = 0$ we have

$$A_n = \begin{cases} \frac{4n}{\pi(n^2-1)} & n \text{ even} \\ 0 & n \text{ odd.} \end{cases}$$

Thus the sine series for $\cos x$ is

$$\sum_{n \text{ even}} \frac{4n}{\pi(n^2-1)} \sin(nx).$$

It converges to $\cos x$ for $0 < x < \pi$, to $-\cos x$ for $-\pi < x < 0$ and to zero for $x = -\pi, 0, \pi$.

5.4.7.

(a) Since ϕ is an odd function, the coefficients of the cosine terms are all zero. The sine coefficients are

$$B_n = \int_{-1}^{1} \phi(x) \sin n\pi x \, dx$$

$$= 2 \int_{0}^{1} (1-x) \sin n\pi x \, dx$$

$$= \left[\frac{-2}{n\pi} (\cos n\pi x)(1-x) \right]_0^1 - \frac{2}{n\pi} \int_0^1 \cos n\pi x \, dx$$

$$= \frac{2}{n\pi}.$$

Thus the Fourier series for ϕ is $\sum_{n=1}^{\infty} \frac{2}{n\pi} \sin n\pi x$.

(b) The first three nonzero terms are $\frac{2}{\pi} \sin \pi x + \frac{1}{\pi} \sin 2\pi x + \frac{2}{3\pi} \sin 3\pi x$.

(c) Since

$$\|\phi\|^2 = 2 \int_0^1 (1-x)^2 \, dx = 2/3$$

is finite, Theorem 5.4.3 implies that the Fourier series converges to ϕ in the L^2 sense.

(d) By Theorem 5.4.4(ii), the series converges pointwise to ϕ except at $x = 0$, where it converges to 0.

(e) The series does not converge uniformly to ϕ, because there is a point (namely $x = 0$) at which it does not converge to $\phi(x)$.

5.4.8.

(a) The sine series converges pointwise to x^3 on $(0, l)$, but not uniformly, since the odd extension is not continuous at $x = \ell$. The series converges in L^2 since x^3 has finite L^2 norm.

5.4.9. Since $f(x)$ and $\cos \frac{n\pi x}{l}$ are $2l$-periodic, integrating by parts gives

$$a_n' = \frac{1}{l} \int_{-l}^{l} f'(x) \cos \frac{n\pi x}{l}\, dx$$

$$= \left[\frac{1}{l} f(x) \cos \frac{n\pi x}{l} \right]_{-l}^{l} + \frac{n\pi}{l} \frac{1}{l} \int_{-l}^{l} f(x) \sin \frac{n\pi x}{l}\, dx$$

$$= \frac{n\pi b_n}{l}.$$

Likewise,

$$b_n' = \frac{1}{l} \int_{-l}^{l} f'(x) \sin \frac{n\pi x}{l}\, dx$$

$$= \left[\frac{1}{l} f(x) \sin \frac{n\pi x}{l} \right]_{-l}^{l} - \frac{n\pi}{l} \frac{1}{l} \int_{-l}^{l} f(x) \cos \frac{n\pi x}{l}\, dx$$

$$= \frac{-n\pi a_n}{l}.$$

5.4.12. By Example 3 in Section 5.1, the coefficients are

$$A_n = (-1)^{n+1} \frac{2l}{n\pi}.$$

The functions $X_n(x) = \sin(n\pi x/l)$ satisfy $\|X_n\|^2 = l/2$, so by Parseval's equality,

$$\|x\|_{L^2}^2 = \sum_{n=1}^{\infty} A_n^2 \|X_n\|_{L^2}^2 = \frac{2l^3}{\pi^2} \sum_{n=1}^{\infty} \frac{1}{n^2}.$$

Since

$$\|x\|_{L^2}^2 = \int_0^l x^2\, dx = \frac{1}{3} l^3$$

it follows that

$$\sum_{n=1}^{\infty} \frac{1}{n^2} = \frac{\pi^2}{6}.$$

5.4.16 By Theorem 5.4.5, the coefficients in the Fourier series minimize the L^2 error. Since $\phi(x) = |x|$ is even, $b_1 = b_2 = 0$,

$$a_0 = \frac{2}{\pi} \int_0^{\pi} x\, dx = \pi,$$

$$a_1 = \frac{2}{\pi} \int_0^{\pi} x \cos x\, dx = -\frac{4}{\pi},$$

and

$$a_2 = \frac{2}{\pi} \int_0^{\pi} x \cos 2x\, dx = 0.$$

So the best approximation of the given form is $f(x) = \frac{\pi}{2} - \frac{4}{\pi} \cos x$.

5.4.19

(a) Differentiating $-X'' = \lambda X$ with respect to λ gives $-X_\lambda'' = \lambda X_\lambda + X$.

(b) Green's second identity (5.3.3), with $X_1 = X_\lambda$ and $X_2 = X$ gives

$$\int_a^b -X_\lambda'' X + X_\lambda X'' \, dx = \left[-X_\lambda' X + X_\lambda X' \right]_a^b,$$

so using the fact that $X'' = -\lambda X$ and $-X_\lambda'' = \lambda X_\lambda + X$, this becomes

$$\int_a^b X^2 \, dx = \left[-X_\lambda' X + X_\lambda X' \right]_a^b.$$

(c) Let $X(x, \lambda) = \sin(\sqrt{\lambda} \, x)$. Then $X_\lambda = \frac{1}{2} \lambda^{-1/2} x \cos(\sqrt{\lambda} \, x)$ and $X' = \sqrt{\lambda} \cos(\sqrt{\lambda} \, x)$, so part (b) applied at $\lambda = m^2 \pi^2 / l^2$ gives

$$\int_0^l \sin^2 \frac{m\pi x}{l} \, dx = \left[-X_\lambda' \left(x, \frac{m^2 \pi^2}{l^2} \right) \sin \frac{m\pi x}{l} + \frac{1}{2} x \cos^2 \frac{m\pi x}{l} \right]_0^l = \frac{l}{2}.$$

Section 5.5

5.5.2 If g is the zero function, then the inequality holds since both sides are zero. Now suppose g is not the zero function, and let $H(t) = \|f + tg\|^2$. Then $H(t) \geq 0$ for all t. Since

$$H(t) = (f + tg, f + tg) = (f, f) + 2(f, g)t + (g, g)t^2$$
$$H'(t) = 2(f, g) + 2(g, g)t,$$

the only critical point of H is $t = -(f, g)/(g, g)$. Thus

$$0 \leq H(-(f, g)/(g, g)) = (f, f) - 2\frac{(f, g)^2}{(g, g)} + \frac{(f, g)^2}{(g, g)} = (f, f) - \frac{(f, g)^2}{(g, g)}.$$

Multiplying both sides by (g, g), this yields $(f, f)(g, g) - (f, g)^2 \geq 0$, so $(f, g)^2 \leq \|f\|^2 \|g\|^2$. Taking the square root of both sides gives $|(f, g)| \leq \|f\| \cdot \|g\|$.

5.5.4 Parts (a) and (b) follow from Exercise 4.3.12.

(c) Let v be an eigenfunction with eigenvalue λ. That is $-v_{xx} = \lambda v$ and $v_x(0) = v_x(l) = \frac{v(l) - v(0)}{l}$. Applying Green's first identity (see Exercise 5.3.12), we have

$$-\int_0^l v_{xx} v \, dx = \int_0^l v_x^2 \, dx - v_x v \Big|_0^l$$

Using the properties of v, this becomes

$$\lambda \int_0^l v^2 \, dx = \int_0^l v_x^2 \, dx - [v(l) - v(0)]^2 / l.$$

By Exercise 5.5.3, the right hand side is nonnegative, from which it follows that $\lambda \geq 0$.

(d) Note that
$$\phi(x) = A + Bx + (\text{orthogonal terms}),$$
where the orthogonal terms are orthogonal to both 1 and x on $(0, l)$. Let
$$\alpha = \int_0^l \phi(x)\, dx, \qquad \text{and} \qquad \beta = \int_0^l x\phi(x)\, dx.$$

Then
$$\alpha = \int_0^l A + Bx\, dx = Al + \frac{1}{2}Bl^2$$

and
$$\beta = \int_0^l x(A + Bx)\, dx = \frac{1}{2}Al^2 + \frac{1}{3}Bl^3.$$

Then we can solve these two equations for A and B. We get
$$A = \frac{4l\alpha - 6\beta}{l^2} \qquad \text{and} \qquad B = \frac{12\beta - 6l\alpha}{l^3}.$$

5.5.6 The solution of the diffusion equation is given by the series

$$u(x, t) = \sum_{n=1}^{\infty} A_n \sin\left(\frac{n\pi x}{l}\right) e^{-n^2 \pi^2 kt/l^2}$$

where

$$A_n = \frac{2}{l} \int_0^l \phi(x) \sin\frac{n\pi x}{l}\, dx.$$

Since ϕ is continuous on $[0, l]$, $|\phi(x)|$ is bounded by some constant M on $[0, l]$. Thus

$$|A_n| \leq \frac{2}{l} \int_0^l M\, dx = 2M$$

for all n. Now let
$$f_n(x, t) = A_n \sin\left(\frac{n\pi x}{l}\right) e^{-n^2 \pi^2 kt/l^2}.$$

Then
$$\frac{\partial f_n}{\partial x} = A_n \frac{n\pi}{l} \cos\left(\frac{n\pi x}{l}\right) e^{-n^2 \pi^2 kt/l^2}.$$

Note that
$$\frac{n\pi}{l} e^{-n^2 \pi^2 kt/l^2} \leq C e^{-n\pi^2 kt/l^2} = C\left[e^{-\pi^2 kt/l^2}\right]^n$$

for some constant C. Fix $t > 0$ and let $r = e^{-\pi^2 kt/l^2}$. Then $0 < r < 1$ and we have

$$\left|\frac{\partial f_n}{\partial x}\right| \leq |A_n| e^{-n^2 \pi^2 kt/l^2} \leq 2MC(e^{-\pi^2 kt/l^2})^n = 2MCr^n$$

for all $x \in [0, l]$ and $t > 0$. Hence the series $\sum_{n=1}^{\infty} \frac{\partial f_n}{\partial x}$ converges uniformly in x by comparison with the geometric series $\sum 2MCr^n$. Since the series $\sum f_n(x, t)$ converges to $u(x, t)$, it follows from the theorem at the end of Appendix A.2 that

$$u_x(x, t) = \sum_{n=1}^{\infty} \frac{\partial f_n}{\partial x},$$

so u is differentiable with respect to x. The same argument shows that

$$u_t(x, t) = \sum_{n=1}^{\infty} \frac{\partial f_n}{\partial t}.$$

In general, any higher-order derivative of f_n takes the form

$$A_n \cdot (\text{polynomial in } n) \cdot (\sin \text{ or } \cos \text{ of } n\pi x/l) \cdot e^{-n^2 \pi^2 kt/l^2},$$

which can be bounded by Cr^n for some constant C. Hence each term-by-term differentiated series converges uniformly.

5.5.7 The coefficients c_n are given by

$$c_n = \frac{1}{2\pi} \int_{-\pi}^{\pi} f(x) e^{-inx} \, dx$$

so since

$$\sum_{n=-N}^{N} e^{-inx} = \frac{e^{iNx} - e^{-i(N+1)x}}{1 - e^{ix}},$$

we have

$$\sum_{n=-N}^{N} c_n = \frac{1}{2\pi} \int_{-\pi}^{\pi} f(x) \frac{e^{-i(N+1)x} - e^{iNx}}{e^{ix} - 1} \, dx.$$

Let $X_n = e^{-inx}$ and let $g(x) = f(x)/(e^{ix} - 1)$. Then the above expression is just

$$\frac{1}{2\pi} \left[(g, X_{N+1}) - (g, X_{-N}) \right].$$

It therefore suffices to show that $(g, X_N) \to 0$ as $N \to \pm\infty$. Since $\|X_N\|^2 = 2\pi$, it follows from Bessel's inequality that

$$\frac{1}{2\pi} \sum_{N=-\infty}^{\infty} (g, X_N)^2 \leq \|g\|^2,$$

which is finite by assumption. Thus the series converges, so $(g, X_N) \to 0$ as $N \to \pm\infty$.

5.5.8 First we need to show that the functions $X_N = \sin(N + \frac{1}{2})\theta$ are orthogonal on $(-\pi, 0)$ and $(0, \pi)$. On $(-\pi, 0)$ they are eigenfunctions corresponding to the boundary conditions $X'(-\pi) = X(0) = 0$, which are symmetric, so Theorem 5.3.1 implies that they are orthogonal

57

on $(-\pi, 0)$. On $(0, \pi)$, they are eigenfunctions with boundary conditions $X(0) = X'(\pi) = 0$, which are also symmetric, so the same Theorem implies they are orthogonal on $(0, \pi)$.

Next, since g_+ is piecewise continuous, its square is also piecewise continuous, and therefore integrable over $(-\pi, 0)$, so the L^2 norm of g_+ is finite. Thus by Bessel's inequality,

$$\sum_{N=0}^{\infty} \frac{(g_+, X_N)}{\|X_N\|^2} \leq \|g_+\|^2.$$

Since $\|X_N\|^2 = \pi/2$, this implies that $(g_+, X_N) \to 0$ as $N \to \infty$. Likewise $(g_-, X_N) \to 0$ as $N \to \infty$. Therefore

$$\int_0^\pi g_+(\theta) \sin(N + \frac{1}{2})\theta \, d\theta + \int_{-\pi}^0 g_-(\theta) \sin(N + \frac{1}{2})\theta \, d\theta = (g_+, X_N) + (g_-, X_N) \to 0$$

as $N \to \infty$.

5.5.8 First we need to show that the functions $X_N = \sin(N + \frac{1}{2})\theta$ are orthogonal on $(-\pi, 0)$ and $(0, \pi)$. On $(-\pi, 0)$ they are eigenfunctions corresponding to the boundary conditions $X'(-\pi) = X(0) = 0$, which are symmetric, so Theorem 5.3.1 implies that they are orthogonal on $(-\pi, 0)$. On $(0, \pi)$, they are eigenfunctions with boundary conditions $X(0) = X'(\pi) = 0$, which are also symmetric, so the same Theorem implies they are orthogonal on $(0, \pi)$.

Next, since g_+ is piecewise continuous, its square is also piecewise continuous, and therefore integrable over $(-\pi, 0)$, so the L^2 norm of g_+ is finite. Thus by Bessel's inequality,

$$\sum_{N=0}^{\infty} \frac{(g_+, X_N)}{\|X_N\|^2} \leq \|g_+\|^2.$$

Since $\|X_N\|^2 = \pi/2$, this implies that $(g_+, X_N) \to 0$ as $N \to \infty$. Likewise $(g_-, X_N) \to 0$ as $N \to \infty$. Therefore

$$\int_0^\pi g_+(\theta) \sin(N + \frac{1}{2})\theta \, d\theta + \int_{-\pi}^0 g_-(\theta) \sin(N + \frac{1}{2})\theta \, d\theta = (g_+, X_N) + (g_-, X_N) \to 0$$

as $N \to \infty$.

5.5.12 Writing the full Fourier series of f as

$$f(x) = \frac{1}{2}a_0 + \sum_{n=1}^{\infty} a_n \cos nx + b_n \sin nx,$$

we have $a_0 = \frac{1}{\pi} \int_{-\pi}^\pi f(x) \, dx = 0$ by assumption. By Parseval's equality (5.4.19),

$$\int_{-\pi}^\pi |f(x)|^2 \, dx = \sum_{n=1}^{\infty} a_n^2 \left(\int_{-\pi}^\pi \cos^2 nx \, dx \right) + b_n^2 \left(\int_{-\pi}^\pi \sin^2 nx \, dx \right)$$

$$= \pi \sum_{n=1}^{\infty} (a_n^2 + b_n^2).$$

Similarly, if we write

$$f'(x) = \frac{1}{2}a_0' + \sum_{n=1}^{\infty} a_n' \cos nx + b_n' \sin nx,$$

then we have $a_0' = \frac{1}{\pi}\int_{\pi}^{\pi} f'(x)\, dx = 0$. By Exercise 5.4.9 with $l = \pi$, we have $a_n' = nb_n$ and $b_n' = -na_n$. Using Parseval's equality again gives

$$\int_{-\pi}^{\pi} |f'(x)|^2 \, dx = \pi \sum_{n=1}^{\infty} \left((nb_n)^2 + (-na_n)^2\right) = \pi \sum_{n=1}^{\infty} n^2(a_n^2 + b_n^2)$$

$$\geq \pi \sum_{n=1}^{\infty} (a_n^2 + b_n^2) = \int_{-\pi}^{\pi} |f(x)|^2 \, dx.$$

Section 5.6

5.6.1. Solution 1. Let $v(x,t) = u(x,t) - 1$. Then v satisfies

$$v_t = v_{xx}$$
$$v_x(0,t) = 0, v(1,t) = 0$$
$$v(x,0) = x^2 - 1$$

The eigenvalues and eigenfunctions for the problem $-X'' - \lambda X$, $X'(0) = 0$, $X(1) = 0$ are

$$\lambda_n = (n + \tfrac{1}{2})^2\pi^2 \qquad \text{and} \qquad X_n(x) = \cos((n + \tfrac{1}{2})\pi x)$$

for $n \geq 0$. Thus

$$v(x,t) = 1 + \sum_{n=0}^{\infty} A_n \cos((n + \tfrac{1}{2})\pi x)e^{-(n+\frac{1}{2})^2\pi^2 t},$$

where

$$A_n = 2\int_0^1 (x^2 - 1)\cos((n + \tfrac{1}{2})\pi x)\, dx = \frac{4(-1)^{n+1}}{(n + \frac{1}{2})^3 \pi^3}.$$

Therefore

$$u(x,t) = 1 + \sum_{n=0}^{\infty} \frac{4(-1)^{n+1}}{(n + \frac{1}{2})^3 \pi^3} \cos((n + \tfrac{1}{2})\pi x)e^{-(n+\frac{1}{2})^2\pi^2 t}.$$

As $t \to \infty$, every term in the series converges to zero, so the only term left is 1.

Solution 2. Let $v(x,t) = u(x,t) - x^2$. Then v satisfies

$$v_t = v_{xx} + 2$$
$$v_x(0,t) = 0, v(1,t) = 0$$
$$v(x,0) = 0$$

Applying the solution operator to the inhomogeneous term we get

$$\mathscr{S}(t-s)(2) = \sum_{n=0}^{\infty} A_n \cos((n+\tfrac{1}{2})\pi x) e^{-(n+\frac{1}{2})^2 \pi^2 (t-s)},$$

where

$$A_n = 2 \int_0^1 2 \cos((n+\tfrac{1}{2})\pi x)\, dx = \frac{4(-1)^n}{(n+\frac{1}{2})\pi}.$$

Using Duhamel's principle, the solution takes the form

$$\begin{aligned}
v(x,t) &= \int_0^t \mathscr{S}(t-s)(2)\, ds \\
&= \sum_{n=0}^{\infty} \frac{4(-1)^n}{(n+\frac{1}{2})\pi} \cos((n+\tfrac{1}{2})\pi x) \int_0^t e^{-(n+\frac{1}{2})^2 \pi^2 (t-s)}\, ds \\
&= \sum_{n=0}^{\infty} \frac{4(-1)^n}{(n+\frac{1}{2})^3 \pi^3} \cos((n+\tfrac{1}{2})\pi x) \left[1 - e^{-(n+\frac{1}{2})^2 \pi^2 t}\right],
\end{aligned}$$

and thus

$$u(x,t) = x^2 + \sum_{n=0}^{\infty} \frac{4(-1)^n}{(n+\frac{1}{2})^3 \pi^3} \cos((n+\tfrac{1}{2})\pi x) \left[1 - e^{-(n+\frac{1}{2})^2 \pi^2 t}\right].$$

In view of the series expansion of $x^2 - 1$ calculated above, this does agree with the solution obtained by the other method.

5.6.4. Solution 1. First note that the function $v = -\frac{k}{2c^2}(x^2 - 2lx)$ satisfies the PDE and boundary condition. Thus $w = u - v$ satisfies

$$\begin{aligned}
w_{tt} &= c^2 w_{xx} \\
w(0,t) &= w_x(l,t) = 0 \\
w(x,0) &= \frac{k}{2c^2}(x^2 - 2lx), \quad w_t(x,0) = V.
\end{aligned}$$

The eigenfunctions of this homogeneous problem are $\sin(\beta_n x)$ where $\beta_n = (n+\tfrac{1}{2})\pi/l$ for $n \geq 0$, so we may write

$$w(x,t) = \sum_{n=1}^{\infty} \sin(\beta_n x) \left[A_n \cos(\beta_n ct) + B_n \sin(\beta_n ct)\right].$$

The initial condition $w(x,0) = \frac{k}{2c^2}(x^2 - 2lx)$ implies

$$A_n = \frac{2}{l} \int_0^l \left(\frac{k}{2c^2}(x^2 - 2lx)\right) \sin(\beta_n x)\, dx = -\frac{2k}{lc^2 \beta_n^3},$$

and the condition $w_t(x,0) = V$ implies

$$B_n = \frac{2}{l} \int_0^l \frac{V}{c\beta_n} \sin(\beta_n x)\, dx = \frac{2V}{lc\beta_n^2}.$$

Thus

$$u(x,t) = -\frac{k}{2c^2}(x^2 - 2lx) + \sum_{n=1}^{\infty} \sin(\beta_n x)\left[-\frac{2k}{lc^2\beta_n^3}\cos(\beta_n ct) + \frac{2V}{lc\beta_n^2}\sin(\beta_n ct)\right].$$

Solution 2. The eigenfunctions of the homogeneous problem are $\sin(\beta_n x)$ where $\beta_n = (n+\frac{1}{2})\pi/l$ for $n \geq 0$. Therefore, we write

$$u(x,t) = \sum_{n=0}^{\infty} u_n(t)\sin(\beta_n x)$$

where

$$u_n(t) = \frac{2}{l}\int_0^l u(x,t)\sin(\beta_n x)\,dx.$$

Next, we write

$$u_{tt} = \sum_{n=0}^{\infty} v_n(t)\sin(\beta_n x)$$

where

$$v_n(t) = \frac{2}{l}\int_0^l u_{tt}(x,t)\sin(\beta_n x)\,dx = \frac{d^2 u_n}{dt^2}.$$

Finally, we write

$$u_{xx} = \sum_{n=0}^{\infty} w_n(t)\sin(\beta_n x)$$

where

$$w_n(t) = \frac{2}{l}\int_0^l u_{xx}(x,t)\sin(\beta_n x)\,dx.$$

Using Green's second identity, we see that

$$w_n(t) = -\frac{2}{l}\beta_n^2\int_0^l u(x,t)\sin(\beta_n x)\,dx + \frac{2}{l}\left[u_x(x,t)\sin(\beta_n x) - \beta_n u(x,t)\cos(\beta_n x)\right]\Big|_{x=0}^{x=l}$$

$$= -\beta_n^2 u_n.$$

By assumption, $u_{tt} - c^2 u_{xx} = k$. Writing

$$k = \sum_{n=0}^{\infty} f_n(t)\sin(\beta_n x),$$

we have

$$f_n(t) = \frac{2}{l}\int_0^l k\sin(\beta_n x)\,dx = \frac{2k}{l\beta_n}.$$

Therefore,

$$u_{tt} - c^2 u_{xx} = \sum_{n=0}^{\infty}[v_n(t) - c^2 w_n(t)]\sin(\beta_n x) = \sum_{n=0}^{\infty} f_n(t)\sin(\beta_n x) = k,$$

so

$$v_n(t) - c^2 w_n(t) = \frac{2k}{l\beta_n}.$$

Using the facts that $v_n(t) = d^2 u_n/dt^2$ and $w_n(t) = -\beta_n^2 u_n$, we have

$$\frac{d^2 u_n}{dt^2} + c^2 \beta_n^2 u_n = \frac{2k}{l\beta_n}.$$

Using the method of undetermined coefficients, we see that a particular solution of this inhomogeneous ODE is given by

$$(u_n)_p = \frac{2k}{lc^2 \beta_n^3}.$$

Therefore, the general solution of this second-order ODE is given by

$$u_n(t) = A\cos(c\beta_n t) + B\sin(c\beta_n t) + \frac{2k}{lc^2 \beta_n^3}.$$

By assumption, $u(x,0) = 0$. Therefore,

$$u_n(0) = A + \frac{2k}{lc^2 \beta_n^3} = 0 \implies A = -\frac{2k}{lc^2 \beta_n^3}.$$

Also, by assumption, $u_t(x,0) = V$. Using the fact that

$$u_t(x,t) = \sum_{n=0}^{\infty} y_n(t)\sin(\beta_n x)$$

together with

$$y_n(t) = \frac{2}{l}\int_0^l u_t(x,t)\sin(\beta_n x)\,dx = \frac{du_n}{dt},$$

we see that

$$\frac{du_n}{dt}(0) = \frac{2}{l}\int_0^l V\sin(\beta_n x)\,dx = \frac{2V}{l\beta_n}.$$

Therefore,

$$u_n'(0) = Bc\beta_n = \frac{2V}{l\beta_n} \implies B = \frac{2V}{lc\beta_n^2}.$$

We conclude that

$$u(x,t) = \sum_{n=0}^{\infty} u_n(t)\sin(\beta_n x)$$

where $\beta_n = (n + \frac{1}{2})\pi/l$ and

$$u_n(t) = -\frac{2k}{lc^2 \beta_n^3}\cos(c\beta_n t) + \frac{2V}{lc\beta_n^2}\sin(c\beta_n t) + \frac{2k}{lc^2 \beta_n^3}.$$

Note that this answer agrees with the answer found using Solution 1, because the term $\frac{2k}{lc^2 \beta_n^3}$ is the nth coefficient of the Fourier series of $-\frac{k}{2c^2}(x^2 - 2lx)$.

5.6.6. The problem takes the form of equation (5.6.11), with $f(x,t) = g(x)\sin(\omega t)$, $h(t) = k(t) = 0$ and $\phi(x) = \psi(x) = 0$. Therefore, writing the solution in the form

$$u(x,t) = \sum_{n=1}^{\infty} u_n(t) \sin\frac{n\pi x}{l},$$

the ODE in equation (5.6.12) that the coefficients u_n must satisfy becomes

$$\frac{d^2 u_n}{dt^2} + \frac{c^2 n^2 \pi^2}{l^2} u_n = A_n \sin\omega t,$$

where

$$A_n = \frac{2}{l} \int_0^l g(x) \sin\frac{n\pi x}{l}\, dx,$$

and the initial conditions are given by $u_n(0) = u_n'(0) = 0$. For $\omega \neq \frac{n\pi c}{l}$, the solution of this ODE is

$$u_n(t) = \frac{A_n}{\frac{n^2 \pi^2 c^2}{l^2} - \omega^2} \left(\sin\omega t - \frac{\omega l}{n\pi c} \sin\frac{n\pi c t}{l} \right)$$

and for $\omega = \frac{n\pi c}{l}$, the solution is

$$u_n(t) = -\frac{A_n}{2\omega} \left(t\cos\omega t - \frac{1}{\omega}\sin\omega t \right).$$

The term $t\cos\omega t$ grows (in amplitude) with time. Therefore resonance occurs when $\omega = \frac{n\pi c}{l}$ for some positive integer n, and the coefficient A_n is nonzero, i.e. when $g(x)$ is not orthogonal to the function $\sin\frac{n\pi x}{l}$.

5.6.9. Let $w(x) = (1-x)h + xk$, and let $v = u - w$. Then v satisfies

$$v_{tt} = 9v_{xx}$$
$$v(0,t) = v(1,t) = 0$$
$$v(x,0) = -w(x),\ v_t(x,0) = 0.$$

The solution of this problem is

$$v(x,t) = \sum_{n=1}^{\infty} \left(A_n \cos(3n\pi t) + B_n \sin(3n\pi t) \right) \sin(n\pi x),$$

where

$$A_n = 2\int_0^1 -w(x)\sin(n\pi x)\, dx \qquad \text{and} \qquad B_n = \frac{2}{n\pi 3} \int_0^1 0\, dx = 0.$$

Integrating by parts reveals that

$$A_n = \frac{2(k(-1)^n - h)}{n\pi},$$

so

$$u(x,t) = (1-x)h + xk + 2\sum_{n=1}^{\infty} \frac{k(-1)^n - h}{n\pi} \cos(3n\pi t)\sin(n\pi x).$$

5.6.10. To derive the PDE, follow Example 1.3.5. Let $H(t) = \iiint_D c\rho u\, dz\, dy\, dx$, where D is the slab of the con between $x = x_1$ and $x = x_2$. Then

$$\frac{dH}{dt} = c\rho \int_{x_1}^{x_2} u_t A(x)\, dx.$$

By the law of heat flow,

$$\frac{dH}{dt} = \iint_{\text{bdy } D} \kappa \mathbf{n} \cdot \nabla u\, dS = \iint_{x=x_2} \kappa[1,0,0] \cdot \nabla u\, dz\, dy + \iint_{x=x_1} \kappa[-1,0,0] \cdot \nabla u\, dz\, dy$$
$$= \kappa A(x_2) u_x(x_2, t) - \kappa A(x_1) u_x(x_1, t).$$

Differentiating with respect to x_2, we get

$$c\rho A(x) u_t = \kappa (A(x) u_x)_x,$$

or

$$u_t = \frac{k}{A(x)} (A(x) u_x)_x, \qquad 0 < x < l,$$

where $k = \frac{\kappa}{c\rho}$. The boundary conditions are $u(0,t) = 0$ and $|u(l,t)| < \infty$. The initial condition is $u(x,0) = \phi(x)$. Now, separating variables $u = X(x)T(t)$, we get

$$\frac{T'(t)}{kT(t)} = \frac{[A(x)X'(x)]'}{A(x)X(x)} = \text{constant} \equiv -\lambda.$$

As usual, $T(t) = Ce^{-k\lambda t}$ and $[AX']' = -\lambda A X$. That is, $\left[\left(1 - \frac{x}{l}\right)^2 X'\right]' = -\lambda \left(1 - \frac{x}{l}\right)^2 X$. To solve this ODE, substitute $X = \left(1 - \frac{x}{l}\right)^{-1} v$. Then

$$X' = \left(1 - \frac{x}{l}\right)^{-1} v' + \frac{1}{l}\left(1 - \frac{x}{l}\right)^{-2} v$$

$$\left(1 - \frac{x}{l}\right)^2 X' = \left(1 - \frac{x}{l}\right) v' + \frac{1}{l} v$$

$$\left[\left(1 - \frac{x}{l}\right)^2 X'\right]' = \left(1 - \frac{x}{l}\right) v'' - \frac{1}{l} v' + \frac{1}{l} v' = \left(1 - \frac{x}{l}\right) v''.$$

Therefore $\left(1 - \frac{x}{l}\right) v'' = -\lambda \left(1 - \frac{x}{l}\right)^2 X = -\lambda \left(1 - \frac{x}{l}\right) v$. Thus $v'' = -\lambda v$. The boundary conditions are $v(0) = 0$ and $v(l) = \lim_{x \to l^-}\left(1 - \frac{x}{l}\right) X(x) = 0$. From this it follows that $v(x) = \sin\frac{n\pi x}{l}$ and $\lambda = \left(\frac{n\pi}{l}\right)^2$. Therefore

$$u(x,t) = \left(1 - \frac{x}{l}\right)^{-1} \sum_{n=1}^{\infty} A_n e^{-n^2\pi^2 kt/l^2} \sin\frac{n\pi x}{l},$$

64

where

$$A_n = \frac{2}{l^2} \int_0^l (l - x)\, \phi(x) \sin \frac{n\pi x}{l}\, dx.$$

5.6.13.

(a) Using the assumption that $\beta^2 c^2 = \omega^2 - ir\omega$ gives

$$\mathcal{U}_{tt} - c^2 \mathcal{U}_{xx} + r\mathcal{U}_t = Ae^{i\omega t} \frac{\sin \beta x}{\sin \beta l} \left(-\omega^2 + c^2\beta^2 + ir\omega \right) = 0,$$

so that $\mathcal{U}$ satisfies the PDE.

(b) Let $v = u - \mathcal{U}$. Then v satisfies the same PDE, but with homogeneous boundary conditions $v(0, t) = v(l, t) = 0$. Writing v in the separated form $v = XT$ and plugging into the PDE, we get

$$\frac{T''}{c^2 T} + r\frac{T'}{c^2 T} = \frac{X''}{X} = \text{constant} = -\lambda.$$

By the boundary conditions, we have $X(0) = X(l) = 0$, so $X_n = \sin \frac{n\pi x}{l}$ and $\lambda_n = \frac{n^2\pi^2}{l^2}$. Therefore T_n satisfies the ODE

$$T_n'' + rT_n' + \lambda_n c^2 T_n = 0.$$

Solutions of this equation take the form $e^{\gamma_n t}$, where

$$\gamma_n = \frac{-r \pm \sqrt{r^2 - 4\lambda_n c^2}}{2}.$$

Since $r > 0$ and $\lambda_n > 0$ for each n, either γ_n is real and negative, or γ_n is complex with negative real part. In either case, it follows that

$$\lim_{t \to \infty} T_n(t) = 0,$$

and thus $v(x, t) \to 0$ as $t \to \infty$. So $u \to \mathcal{U}$ as $t \to \infty$.

(c) When $r = 0$, the equation takes the form of (5.6.11) with $f(x, t) = 0$, $h(t) = 0$ and $k(t) = Ae^{i\omega t}$. Assuming a solution of the form

$$u(x, t) = \sum_{n=1}^{\infty} u_n(t) \sin \frac{n\pi x}{l},$$

the coefficients satisfy equation (5.6.12) in the form

$$\frac{d^2 u_n}{dt^2} + \frac{c^2 n^2 \pi^2}{l^2} u_n = -\frac{2n\pi(-1)^n}{l^2} Ae^{i\omega t}.$$

with initial conditions $u_n(0) = \phi_n$ and $u_n'(0) = \psi_n$. If $\omega = \frac{m\pi c}{l}$, the solution of the ODE is

$$u_m(t) = \frac{\pi(-1)^m iA}{l^2 \omega} te^{i\omega t} + a_m \cos \omega t + b_m \sin \omega t,$$

65

where a_n and b_n are determined by ϕ_n and ψ_n. The first term grows in amplitude as t increases, so we have resonance.

(d) Part (c) shows that there is resonance when $r = 0$. Since $\mathscr{U}(x, t)$ is bounded, part (b) shows that there is no resonance when $r > 0$. Therefore friction prevents resonance.

Chapter 6

Section 6.1

6.1.2. If $u(x, y, z)$ depends only on r, then $\Delta u = u_{rr} + \frac{2}{r}u_r$ by equation (6.1.6). Letting $v = ru$ gives $v_r = ru_r + u$ and $v_{rr} = ru_{rr} + 2u_r$. Hence the equation $\Delta u = k^2 u$ reduces to $v_{rr} = k^2 v$, which has the solutions $v = A\cosh(kr) + B\sinh(kr)$. Therefore $u = \frac{1}{r}[A\cosh(kr) + B\sinh(kr)]$.

6.1.4. Since radial solutions of Laplace's equation in three dimensions are given by $c_1 r^{-1} + c_2$, we have $c_1/a + c_2 = A$ and $c_1/b + c_2 = B$. Solving gives $c_1 = (A - B)/(1/a - 1/b)$ and $c_2 = (A/b - B/a)/(1/b - 1/a)$. Thus

$$u = \frac{ab}{a - b}\left(\frac{B - A}{r} + \frac{A}{b} - \frac{B}{a}\right).$$

6.1.6. Assume a radial solution, so that $u_{rr} + \frac{1}{r}u_r = 1$, and thus $(ru_r)_r = r$. Integrating and dividing by r gives $u_r = \frac{1}{2}r + c_1 r^{-1}$. Integrating again gives $u = \frac{1}{4}r^2 + c_1 \ln r + c_2$. The boundary conditions $u(a) = u(b) = 0$ imply $c_1 = \frac{b^2 - a^2}{4(\ln a - \ln b)}$ and $c_2 = \frac{a^2 \ln b - b^2 \ln a}{4(\ln a - \ln b)}$.

6.1.7. Assume a radial solution, so that $u_{rr} + \frac{2}{r}u_r = 1$, and thus $(r^2 u_r)_r = r^2$. Integrating and dividing by r^2 gives $u_r = \frac{1}{3}r + c_1 r^{-2}$. Integrating again gives $u = \frac{1}{6}r^2 - c_1 r^{-1} + c_2$. The boundary conditions imply $\frac{1}{6}a^2 - c_1/a + c_2 = 0$ and $\frac{1}{6}b^2 - c_1/b + c_2 = 0$. Thus $c_1 = \frac{1}{6}(a^2 - b^2)/(1/a - 1/b) = -\frac{1}{6}ab(a + b)$ and $c_2 = -\frac{1}{6}(a^2 + ab + b^2)$. Thus

$$u = \frac{1}{6}\left(r^2 + \frac{ab(a + b)}{r} - (a^2 + ab + b^2)\right).$$

6.1.9.

(a) A steady state solution of the heat equation satisfies Laplace's equation, $\Delta u = 0$. Since the boundary conditions depend only on the radius, we may seek a solution u which depends only on r. Using equation (6.1.6), Laplace's equation becomes $u_{rr} + \frac{2}{r}u_r = 0$, and the solution of this ODE is $u(r) = \frac{c_1}{r} + c_2$. The boundary condition $u_r(2) = -\gamma$ implies $c_1 = 4\gamma$, and the boundary condition $u(1) = 100$ implies $c_1 + c_2 = 100$, so $c_2 = 100 - 4\gamma$. Therefore $u = 100 - 4\gamma + \frac{4\gamma}{r}$.

(b) The solution in part (a) is decreasing in r, so the hottest temperature is $u(1) = 100$ and the coldest temperature is $u(2) = 100 - 2\gamma$.

(c) Yes, choose $\gamma = 40$.

6.1.11. Any solution must satisfy

$$\iiint_D f\, d\mathbf{x} = \iiint_D \Delta u\, d\mathbf{x} = \iiint_D \operatorname{div}\nabla u\, d\mathbf{x} = \iint_{\text{bdy } D} \frac{\partial u}{\partial n}\, dS = \iint_{\text{bdy } D} g\, dS.$$

Section 6.2

6.2.1. Guess that $u = Ax^2 + Bxy + Cy^2 + Dx + Ey + F$. Then $u_{xx} + u_{yy} = 2A + 2C$. If u is harmonic, then $C = -A$. The boundary conditions lead to the equations

$$By + D = -a$$
$$Bx + E = b$$
$$2Aa + By + D = 0$$
$$By + 2Cb + E = 0.$$

The first two equations imply $B = 0$, $D = -a$ and $E = b$. Plugging these results into the last two equations gives $A = \frac{1}{2}$ and $C = -\frac{1}{2}$. Thus $u(x,y) = \frac{1}{2}(x^2 - y^2) - ax + by + F$, where F is arbitrary.

6.2.3. Separating variables gives $\frac{X''}{X} + \frac{Y''}{Y} = 0$, so $X'' = \lambda X$ and $Y'' = -\lambda Y$ for some constant λ. Since Y satisfies the boundary conditions $Y'(0) = Y'(\pi) = 0$, we have $Y_n = \cos ny$ and $\lambda_n = n^2$. The solutions of $X'' = n^2 X$ with boundary condition $X(0) = 0$ are $X_n = \sinh nx$ for $n > 0$ and $X_0 = x$, so we have the expansion

$$u(x,y) = A_0 x + \sum_{n=1}^{\infty} A_n \sinh nx \cos ny.$$

The boundary condition $u(\pi, y) = \frac{1}{2}(1 + \cos 2y)$ then implies

$$\frac{1}{2}(1 + \cos 2y) = A_0 \pi + \sum_{n=1}^{\infty} A_n \sinh n\pi \cos ny.$$

Hence $A_0 = \frac{1}{2\pi}$, $A_2 = \frac{1}{2\sinh 2\pi}$ and $A_n = 0$ for all other n, so the solution is $u(x,y) = \frac{1}{2\pi} x + \frac{1}{2\sinh 2\pi} \sinh 2x \cos 2y$.

6.2.6. Separating variables gives

$$\frac{X''}{X} + \frac{Y''}{Y} + \frac{Z''}{Z} = 0.$$

Each of the three terms is constant, so

$$X'' = -\lambda X$$
$$Y'' = -\mu Y$$
$$Z'' = (\lambda + \mu)Z.$$

The boundary conditions imply $X'(0) = X'(1) = Y'(0) = Y'(1) = Z'(0) = 0$. Thus $X_n(x) = \cos n\pi x$ and $\lambda_n = n^2 \pi^2$ for $n = 0, 1, 2, \ldots$; $Y_m(y) = \cos m\pi y$ and $\mu_m = m^2 \pi^2$ for $m = 0, 1, 2, \ldots$; and $Z_{mn}(z) = \cosh(\sqrt{m^2 + n^2}\pi z)$. Writing the solution as a series

$$u(x,y,z) = \sum_{m=0}^{\infty} \sum_{n=0}^{\infty} A_{mn} \cos n\pi x \cos m\pi y \cosh(\sqrt{m^2 + n^2}\pi z),$$

the last boundary condition implies

$$g(x,y) = \sum_{m=0}^{\infty} \sum_{n=0}^{\infty} A_{mn} \pi \sqrt{m^2 + n^2} \cos n\pi x \cos m\pi y \sinh(\sqrt{m^2 + n^2}\pi).$$

The coefficients are therefore given by

$$A_{mn} = \frac{4}{\pi\sqrt{m^2+n^2}\sinh(\sqrt{m^2+n^2}\pi)} \int_0^1 \int_0^1 g(x,y) \cos n\pi x \cos m\pi y\, dx dy$$

if m and n are both nonzero,

$$A_{m0} = \frac{2}{\pi m \sinh(m\pi)} \int_0^1 \int_0^1 g(x,y) \cos m\pi y\, dx dy$$

for $m \neq 0$, and

$$A_{0n} = \frac{2}{\pi n \sinh(n\pi)} \int_0^1 \int_0^1 g(x,y) \cos n\pi x\, dx dy$$

for $n \neq 0$. The constant coefficient A_{00} is arbitrary.

6.2.7.

(a) Separating variables leads to $X'' = -\lambda X$ and $Y'' = \lambda Y$ for some constant λ. The boundary conditions $X(0) = X(\pi) = 0$ imply that $X_n(x) = \sin nx$ and $\lambda_n = n^2$. The solutions of $Y'' = n^2 Y$ are $Y_n = A_n e^{-ny} + B_n e^{ny}$. The boundary condition $\lim_{y\to\infty} Y(y) = 0$ implies that $B_n = 0$. Thus the solution takes the form

$$u(x,y) = \sum_{n=1}^{\infty} A_n \sin(nx)\, e^{-ny}.$$

The boundary condition $u(x,0) = h(x)$ implies

$$h(x) = \sum_{n=1}^{\infty} A_n \sin nx,$$

so the coefficients are given by

$$A_n = \frac{2}{\pi} \int_0^{\pi} h(x) \sin nx\, dx.$$

(b) Without the condition at infinity, solutions would not be unique. They would take the form

$$u(x,y) = \sum_{n=1}^{\infty} \sin nx \left(A_n e^{-ny} + B_n e^{ny} \right),$$

where

$$A_n + B_n = \frac{2}{\pi} \int_0^{\pi} h(x) \sin nx\, dx.$$

Section 6.3

6.3.1.

(a) By the Maximum Principle, u attains its maximum on the boundary. Since the maximum of $3 \sin 2\theta + 1$ is 4, the maximum of u on $\overline{D}$ is 4.

(b) By the Mean Value Property, the value of u at the origin is the average of $3 \sin 2\theta + 1$ on the circumference. So

$$u(0,0) = \frac{1}{2\pi} \int_0^{2\pi} 3 \sin 2\theta + 1 \, d\theta = 1.$$

6.3.2. In the full Fourier series for $h(\theta) = 1 + 3 \sin \theta$, $A_0 = 2$, $B_1 = 3/a$ and all other coefficients are zero. Thus by equation (6.3.10), $u(r,\theta) = 1 + \frac{3}{a} r \sin \theta$. In rectangular coordinates $u(x,y) = 1 + 3y/a$.

Section 6.4

6.4.1. The solution takes the form of equation (6.4.7). The condition at infinity implies $C_n = 0$ for $n \geq 1$ and $D_0 = 0$. Absorbing the coefficients D_n into A_n and B_n gives

$$u(r,\theta) = \frac{1}{2}C_0 + \sum_{n=1}^{\infty} r^{-n}(A_n \cos n\theta + B_n \sin n\theta).$$

The condition at $r = a$ implies

$$1 + 3 \sin \theta = \frac{1}{2}C_0 + \sum_{n=1}^{\infty} a^{-n}(A_n \cos n\theta + B_n \sin n\theta).$$

Thus $C_0 = 2$, $A_n = 0$ for all n, $B_1 = 3a$ and $B_n = 0$ for all other n. Hence $u(r,\theta) = 1 + \frac{3a}{r} \sin \theta$. In rectangular coordinates, $u(x,y) = 1 + \frac{3ay}{x^2+y^2}$.

6.4.3. The solution takes the form

$$u(r,\theta) = \frac{1}{2}(C_0 + D_0 \log r) + \sum_{n=1}^{\infty}(C_n r^n + D_n r^{-n}) \cos n\theta + (A_n r^n + B_n r^{-n}) \sin n\theta.$$

At $r = a$ and $r = b$ we have

$$g(\theta) = \frac{1}{2}(C_0 + D_0 \log a) + \sum_{n=1}^{\infty}(C_n a^n + D_n a^{-n}) \cos n\theta + (A_n a^n + B_n a^{-n}) \sin n\theta$$

$$h(\theta) = \frac{1}{2}(C_0 + D_0 \log b) + \sum_{n=1}^{\infty}(C_n b^n + D_n b^{-n}) \cos n\theta + (A_n b^n + B_n b^{-n}) \sin n\theta.$$

Thus, for $n \geq 1$,

$$A_n a^n + B_n a^{-n} = \frac{1}{\pi} \int_0^{2\pi} g(\theta) \sin n\theta \, d\theta, \qquad C_n a^n + D_n a^{-n} = \frac{1}{\pi} \int_0^{2\pi} g(\theta) \cos n\theta \, d\theta$$

$$A_n b^n + B_n b^{-n} = \frac{1}{\pi} \int_0^{2\pi} h(\theta) \sin n\theta \, d\theta, \qquad C_n b^n + D_n b^{-n} = \frac{1}{\pi} \int_0^{2\pi} h(\theta) \cos n\theta \, d\theta.$$

Solving this system gives

$$A_n = \frac{a^n b^n}{\pi(b^{2n} - a^{2n})} \int_0^{2\pi} \left(a^{-n} h(\theta) - b^{-n} g(\theta) \right) \sin n\theta \, d\theta$$

$$B_n = \frac{a^n b^n}{\pi(b^{2n} - a^{2n})} \int_0^{2\pi} \left(b^n g(\theta) - a^n h(\theta) \right) \sin n\theta \, d\theta$$

$$C_n = \frac{a^n b^n}{\pi(b^{2n} - a^{2n})} \int_0^{2\pi} \left(a^{-n} h(\theta) - b^{-n} g(\theta) \right) \cos n\theta \, d\theta$$

$$D_n = \frac{a^n b^n}{\pi(b^{2n} - a^{2n})} \int_0^{2\pi} \left(b^n g(\theta) - a^n h(\theta) \right) \cos n\theta \, d\theta$$

for $n \geq 1$. For $n = 0$,

$$C_0 + D_0 \log a = \frac{1}{\pi} \int_0^{2\pi} g(\theta) \, d\theta \quad \text{and} \quad C_0 + D_0 \log b = \frac{1}{\pi} \int_0^{2\pi} h(\theta) \, d\theta,$$

so

$$C_0 = \frac{1}{\pi(\log b - \log a)} \int_0^{2\pi} [g(\theta) \log b - h(\theta) \log a] \, d\theta$$

$$D_0 = \frac{1}{\pi(\log b - \log a)} \int_0^{2\pi} [h(\theta) - g(\theta)] \, d\theta.$$

6.4.5.

(a) By formula (6.4.7), the solution takes the form

$$u(r, \theta) = \frac{1}{2}(C_0 + D_0 \log r) + \sum_{n=1}^{\infty} (C_n r^n + D_n r^{-n}) \cos n\theta + (A_n r^n + B_n r^{-n}) \sin n\theta.$$

Insulation at $r = 2$ implies $u_r(2, \theta) = 0$, and thus

$$0 = \frac{1}{4} D_0 + \sum_{n=1}^{\infty} (nC_n 2^{n-1} - nD_n 2^{-n-1}) \cos n\theta + (nA_n 2^{n-1} - nB_n 2^{-n-1}) \sin n\theta.$$

Hence $D_0 = 0$, $D_n = 2^{2n} C_n$ and $B_n = 2^{2n} A_n$, so the solution now takes the form

$$u(r, \theta) = \frac{1}{2} C_0 + \sum_{n=1}^{\infty} (r^n + (4/r)^n)(C_n \cos n\theta + A_n \sin n\theta).$$

71

Now $\sin^2\theta = \frac{1}{2} - \frac{1}{2}\cos 2\theta$, so the condition $u(1,\theta) = \sin^2\theta$ implies

$$\frac{1}{2} - \frac{1}{2}\cos 2\theta = \frac{1}{2}C_0 + \sum_{n=1}^{\infty}(4^n + 1)(C_n\cos n\theta + A_n\sin n\theta).$$

Thus $C_0 = 1$, $C_2 = -1/34$, and all other coefficients are zero. The solution is therefore

$$u(r,\theta) = \frac{1}{2} - \frac{1}{34}(r^2 + 16r^{-2})\cos 2\theta.$$

(b) Beginning with the same form as above, the condition $u(2,\theta) = 0$ implies

$$0 = \frac{1}{2}(C_0 + D_0\log 2) + \sum_{n=1}^{\infty}(C_n 2^n + D_n 2^{-n})\cos n\theta + (A_n 2^n + B_n 2^{-n})\sin n\theta.$$

Therefore $D_n = -4^n C_n$, $B_n = -4^n A_n$ and $C_0 + D_0\log 2 = 0$. Thus the solution takes the form

$$u(r,\theta) = \frac{1}{2}D_0(\log r - \log 2) + \sum_{n=1}^{\infty}(r^n - (4/r)^n)(C_n\cos n\theta + A_n\sin n\theta).$$

Finally, the condition $u(1,\theta) = \sin^2\theta$ implies

$$\frac{1}{2} - \frac{1}{2}\cos 2\theta = \frac{1}{2}D_0(-\log 2) + \sum_{n=1}^{\infty}(1 - 4^n)(C_n\cos n\theta + A_n\sin n\theta).$$

Hence $D_0 = -1/(\log 2)$, $C_2 = 1/30$ and all other coefficients are zero. Thus

$$u(r,\theta) = \frac{1}{2}(1 - \log r/\log 2) + \frac{1}{30}(r^2 - 16r^{-2})\cos 2\theta.$$

6.4.6. By equation (6.4.5) (with $\beta = \pi$), $u(r,\theta) = \sum_{n=1}^{\infty} A_n r^n \sin n\theta$. The boundary condition on the circumference implies $A_1 = \pi$, $A_2 = -1$ and $A_n = 0$ for all other n. Thus $u(r,\theta) = \pi r\sin\theta - r^2\sin 2\theta$. In rectangular coordinates, $u(x,y) = \pi y - 2xy$.

6.4.7 Using the separation of variables technique, we first need to solve the eigenvalue problem,

$$\begin{cases} \Theta'' + \lambda\Theta = 0 \\ \Theta(0) = \Theta(\beta) = 0. \end{cases}$$

The eigenvalues and eigenfunctions are given by

$$\lambda_n = \left(\frac{n\pi}{\beta}\right)^2 \qquad \Theta(\theta) = \sin\left(\frac{n\pi\theta}{\beta}\right).$$

72

The radial equation

$$r^2 R'' + rR' - \lambda R = 0$$

has solutions

$$R_n(r) = C_n r^{n\pi/\beta} + D_n r^{-n\pi/\beta} \qquad n = 1, 2, \ldots$$

As we do not want $u \to +\infty$ as $r \to 0$, we don't allow the solutions $r^{-n\pi/\beta}$. Therefore, we let

$$u(r, \theta) = \sum_{n=1}^{\infty} C_n r^{n\pi/\beta} \sin\left(\frac{n\pi\theta}{\beta}\right).$$

As we want $u(a, \theta) = h(\theta)$, we choose our coefficients C_n as follows.

$$u(a, \theta) = \sum_{n=1}^{\infty} C_n a^{n\pi/\beta} \sin\left(\frac{n\pi\theta}{\beta}\right) = h(\theta).$$

Therefore,

$$C_n = \frac{2}{\beta a^{n\pi/\beta}} \int_0^\beta h(\theta) \sin\left(\frac{n\pi}{\beta}\theta\right) d\theta.$$

6.4.10. By equation (6.4.5) (with $\beta = \pi/2$), $u(r, \theta) = \sum_{n=1}^{\infty} A_n r^{2n} \sin 2n\theta$. By equation (6.4.6), the coefficients are

$$A_n = \frac{2a^{1-2n}}{n\pi} \int_0^{\pi/2} \sin 2n\theta \, d\theta = \begin{cases} 0 & n \text{ even} \\ \frac{2a^{1-2n}}{n^2\pi} & n \text{ odd.} \end{cases}$$

6.4.12.

(a) Let $u(\mathbf{x})$ be a non-constant harmonic function in D, where D is a connected domain with smooth boundary. Suppose $u(\mathbf{x})$ attains its maximum M at $\mathbf{x}_0$ in bdy D. Let B be a disk that lies inside of D and is tangent to bdy D at $\mathbf{x}_0$. The normal vector $\mathbf{n}$ at $\mathbf{x}_0$ is the same for B and for D. So we may as well assume that $D = B$. Suppose that D is the ball $\{|\mathbf{x}| < R\}$ for instance. Let $v(\mathbf{x}) = e^{-\alpha r^2} - e^{-\alpha R^2}$, where $r = |\mathbf{x}|$. Then

$$v_r = -2\alpha r e^{-\alpha r^2}$$
$$v_{rr} = 2\alpha(2\alpha r^2 - 1)e^{-\alpha r^2}$$
$$\Delta v = v_{rr} + \frac{1}{r}v_r = 4\alpha(\alpha r^2 - 1)e^{-\alpha r^2}.$$

Now let A be the annulus $A = \left\{\frac{R}{2} < |\mathbf{x}| < R\right\}$. In A we have

$$\Delta v \geq 4\alpha\left(\alpha\frac{R^2}{4} - 1\right)e^{-\alpha R^2} > 0,$$

provided we choose $\alpha > \frac{4}{R^2}$. Next let $w(\mathbf{x}) = u(\mathbf{x}) - M + \epsilon v(\mathbf{x})$. Then

$$\Delta w = \Delta u + \epsilon\Delta v \geq 0$$

73

in A. Thus w is subharmonic in A. Applying the maximum principle to w in A, we have

$$\max_A w \le \max_{\text{bdy } A} w.$$

On the outer boundary $\{r = R\}$, we have

$$w = u - M + 0 \le 0.$$

On the inner boundary $\{r = \frac{R}{2}\}$ we have

$$w \le \max_{r=R/2} u - M + \epsilon < 0,$$

provided ϵ is chosen small enough. This is because $\max_{r=R/2} u < M$ according to the strong form of the maximum principle. Thus $w(\mathbf{x}) \le 0$ for all $\mathbf{x} \in \text{bdy } A$. Therefore $\max_A w \le 0$. But $w(\mathbf{x_0}) = u(\mathbf{x_0}) - M + \epsilon v(\mathbf{x_0}) = 0 + \epsilon 0 = 0$. Therefore $\frac{\partial w}{\partial n}(\mathbf{x_0}) \ge 0$, so that

$$\frac{\partial u}{\partial n}(\mathbf{x_0}) \ge -\epsilon \frac{\partial v}{\partial n}(\mathbf{x_0}) = 2\epsilon \alpha R e^{-\epsilon R^2} > 0.$$

(b) Suppose $\Delta u = \Delta v = 0$ in D, with $\frac{\partial u}{\partial n} = \frac{\partial v}{\partial n} = h$ on bdy D. Then $w = u - v$ satisfies $\Delta w = 0$ in D with $\frac{\partial w}{\partial n} = 0$ on bdy D. Suppose w is not a constant function. By the strong maximum principle, w attains its maximum at some point $\mathbf{x_0}$ on bdy D. By (a), $\frac{\partial w}{\partial n}(\mathbf{x_0}) > 0$, a contradiction. Therefore w is constant on the connected domain D.

Chapter 7

Section 7.1

7.1.2. Suppose D is a bounded domain and u and v satisfy $\Delta u = \Delta v = f$ on D and $\frac{\partial u}{\partial n} = \frac{\partial v}{\partial n} = g$ on bdy D. Then $w = u - v$ is harmonic on D and $\frac{\partial w}{\partial n} = 0$ on bdy D. By Green's First Identity

$$\iint_{\text{bdy } D} w \frac{\partial w}{\partial n} \, dS = \iiint_D \nabla w \cdot \nabla w \, d\mathbf{x} + \iiint_D w \Delta w \, d\mathbf{x}.$$

By the boundary condition and the fact that w is harmonic, this reduces to

$$\iiint_D |\nabla w|^2 \, d\mathbf{x} = 0,$$

so by the Vanishing Theorem, $|\nabla w(x, y, z)|^2 = 0$ for all $(x, y, z) \in D$. Thus w is constant, so solutions are unique up to constants.

7.1.5. Suppose $E[u] \leq E[w]$ for all functions w on D. Let v be any function on D. Then

$$\mathscr{E}(\epsilon) = E[u + \epsilon v] = \frac{1}{2} \iiint_D |\nabla u + \epsilon \nabla v|^2 \, d\mathbf{x} - \iint_{\text{bdy } D} h(u + \epsilon v) \, dS$$

has a local minimum at $\epsilon = 0$, and therefore

$$0 = \mathscr{E}'(0) = \iiint_D \nabla u \cdot \nabla v \, d\mathbf{x} - \iint_{\text{bdy } D} hv \, dS = \iint_{\text{bdy } D} \left(\frac{\partial u}{\partial n} - h \right) v \, dS - \iiint_D v \Delta u \, d\mathbf{x}.$$

Now let D' be any strict subdomain of D and let $v = 1$ on D' on $v = 0$ on $D - D'$. Then

$$\iiint_{D'} \Delta u \, d\mathbf{x} = 0,$$

and since this holds for all D', the Second Vanishing Theorem implies $\Delta u = 0$ on D. This then implies

$$\iint_{\text{bdy } D} \left(\frac{\partial u}{\partial n} - h \right) v \, dS = 0$$

for all functions v. Choosing v to be a function that is equal to $\frac{\partial u}{\partial n} - h$ on bdy D gives

$$\iint_{\text{bdy } D} \left(\frac{\partial u}{\partial n} - h \right)^2 dS = 0,$$

so $\frac{\partial u}{\partial n} = h$ on bdy D.

7.1.7. Let $w = w_0 + c_1 w_1 + \cdots + c_n w_n$, and define

$$F(c_1, \ldots, c_n) = E(w) = \frac{1}{2} \iiint_D |\nabla w_0 + c_1 \nabla w_1 + \cdots + c_n \nabla w_n|^2 \, d\mathbf{x}.$$

For the choice of coefficients which minimizes F, the partial of F with respect to each c_j must vanish. Hence

$$0 = \frac{\partial F}{\partial c_j} = \iiint_D (\nabla w_0 + c_1 \nabla w_1 + \cdots + c_n \nabla w_n) \cdot \nabla w_j \, d\mathbf{x},$$

or in terms of the inner product,

$$(\nabla w_0, \nabla w_j) + \sum_{k=1}^{n} c_k (\nabla w_k, \nabla w_j) = 0.$$

By symmetry of the inner product this proves the desired result.

7.1.8. By the result of Exercise 7.1.7 with $n = 1$, we have $c_1(\nabla w_1, \nabla w_1) = -(\nabla w_0, \nabla w_1)$. Since $\nabla w_0 = (-3y, 3 - 3x - 2y)$ and $\nabla w_1 = (y(3 - y - 6x), x(3 - 3x - 2y))$, we have

$$(\nabla w_0, \nabla w_1) = \int_0^1 \int_0^{3-3x} [-3y^2(3 - y - 6x) + x(3 - 3x - 2y)^2] \, dy \, dx = \frac{9}{20}.$$

A shortcut to calculate the same integral is to write

$$(\nabla w_0, \nabla w_1) = \iint_D \nabla w_0 \cdot \nabla w_1 \, dy \, dy - \iint_D w_1 \Delta w_0 \, dy \, dx = 2 \iint_D w_1 \, dy \, dx = \frac{9}{20}.$$

Next,

$$(\nabla w_1, \nabla w_1) = \int_0^1 \int_0^{3-3x} [y^2(3 - y - 6x)^2 + x^2(3 - 3x - 2y)^2] \, dy \, dx = \frac{3}{2}.$$

It then follows that $c_1 = -\frac{3}{10}$.

7.1.10. Since $w_0 = x^2$ satisfies the given boundary condition and $w_1 = 1 - x^2 - y^2$ vanishes on the boundary, Exercise 7.1.7 implies that $c_1(\nabla w_1, \nabla w_1) = -(\nabla w_0, \nabla w_1)$. Since $\nabla w_0 = (2x, 0)$ and $\nabla w_1 = (-2x, -2y)$, we have

$$(\nabla w_0, \nabla w_1) = \iint_D -4x^2 \, dA = \int_0^{2\pi} \int_0^1 -4r^3 \cos^2 \theta \, dr \, d\theta = -\pi$$

and

$$(\nabla w_1, \nabla w_1) = \iint_D 4x^2 + 4y^2 \, dA = \int_0^{2\pi} \int_0^1 4r^3 \, dr \, d\theta = 2\pi.$$

Therefore $c_1 = \frac{1}{2}$.

Section 7.2

7.2.1. Let $v = \frac{1}{2\pi} \log r$, where $r = |\mathbf{x} - \mathbf{x}_0|$ and let $D_\epsilon = D - B_\epsilon$, where B_ϵ is the open disk of radius ϵ centered at $\mathbf{x}_0$. Since both u and v are harmonic in D_ϵ, Green's Second Identity implies

$$\int_{\text{bdy } D_\epsilon} \left[u \frac{\partial}{\partial n} \left(\frac{\log r}{2\pi} \right) - \frac{\partial u}{\partial n} \frac{\log r}{2\pi} \right] dS = 0.$$

On the boundary of the disk, $\frac{\partial}{\partial n} = -\frac{\partial}{\partial r}$, so

$$\int_{\text{bdy } D} \left[u \frac{\partial}{\partial n} \left(\frac{\log r}{2\pi} \right) - \frac{\partial u}{\partial n} \frac{\log r}{2\pi} \right] dS = \int_{r=\epsilon} \left[u \frac{\partial}{\partial r} \left(\frac{\log r}{2\pi} \right) - \frac{\partial u}{\partial r} \frac{\log r}{2\pi} \right] dS.$$

The right side equals

$$\frac{1}{2\pi\epsilon} \int_{r=\epsilon} u \, dS - \frac{\epsilon \log \epsilon}{2\pi\epsilon} \int_{r=\epsilon} \frac{\partial u}{\partial r} \, dS = \overline{u} - \epsilon \log \epsilon \overline{u_r}$$

where $\overline{u}$ and $\overline{u_r}$ are the averages over $r = \epsilon$ of u and u_r, respectively. As $\epsilon \to 0$, $\epsilon \log \epsilon \to 0$ and $\overline{u} \to u(\mathbf{x}_0)$, so this implies

$$u(\mathbf{x}_0) = \int_{\text{bdy } D} \left[u \frac{\partial}{\partial n} \left(\frac{\log r}{2\pi} \right) - \frac{\partial u}{\partial n} \frac{\log r}{2\pi} \right] dS$$

$$= \int_{\text{bdy } D} \left[u(\mathbf{x}) \frac{\partial}{\partial n} \left(\frac{\log |\mathbf{x} - \mathbf{x}_0|}{2\pi} \right) - \frac{\partial u}{\partial n}(\mathbf{x}) \frac{\log |\mathbf{x} - \mathbf{x}_0|}{2\pi} \right] dS.$$

7.2.2. Suppose ϕ vanishes outside the ball $r < a$, and let $b > a$. Then ϕ and $\frac{\partial \phi}{\partial n}$ both vanish on the boundary of the ball B_b of radius b centered at the origin. Let $v = -1/(4\pi r)$ and let $D_\epsilon = B_b - B_\epsilon$, where B_ϵ is the ball of radius ϵ centered at the origin. Since v is harmonic on D_ϵ, we have

$$\iiint_{D_\epsilon} v \Delta \phi \, d\mathbf{x} = -\iint_{\text{bdy } D_\epsilon} \left[\phi \frac{\partial v}{\partial n} - v \frac{\partial \phi}{\partial n} \right] dS = -\iint_{r=\epsilon} \left[\phi \frac{\partial v}{\partial n} - v \frac{\partial \phi}{\partial n} \right] dS. \qquad \text{(S-1)}$$

by Green's Second Identity. Next let's take the limit of both sides as $\epsilon \to 0$. For the left hand side, first write

$$\iiint_{D_\epsilon} v \Delta \phi \, d\mathbf{x} = \iiint_{B_b} v \Delta \phi \, d\mathbf{x} - \iiint_{B_\epsilon} v \Delta \phi \, d\mathbf{x}.$$

Since ϕ is assumed to be C^2, $|\Delta \phi| \leq C$ on B_ϵ for some constant C and therefore

$$\left| \iiint_{B_\epsilon} v \Delta \phi \, d\mathbf{x} \right| \leq C \iiint_{B_\epsilon} \frac{1}{4\pi r} \, d\mathbf{x} = C \int_0^{2\pi} \int_0^\pi \int_0^\epsilon \frac{r}{4\pi} \sin\theta \, dr \, d\theta \, d\phi = \frac{1}{2} C \epsilon^2.$$

Hence the left hand side of (S-1) converges to $\iiint_{B_b} v \Delta \phi \, d\mathbf{x}$ as $\epsilon \to 0$. As shown in the proof of equation (7.2.1), the right hand side of (S-1) goes to $\phi(\mathbf{0})$ as $\epsilon \to 0$.

Section 7.3

7.3.1. Suppose G_1 and G_2 are both Green's functions for the domain D at the point $\mathbf{x}_0$. Then by Property (i), $w(\mathbf{x}) = G_1(\mathbf{x}) - G_2(\mathbf{x})$ satisfies $\Delta w = 0$ except possibly at $\mathbf{x}_0$. However, by Property (iii),

$$w(\mathbf{x}) = [G_1(\mathbf{x}) + 1/(4\pi|\mathbf{x} - \mathbf{x}_0|)] - [G_2(\mathbf{x}) + 1/(4\pi|\mathbf{x} - \mathbf{x}_0|)] = H_1(\mathbf{x}) + H_2(\mathbf{x}),$$

where $\Delta H_1(\mathbf{x}_0) = \Delta H_2(\mathbf{x}_0) = 0$, so w is harmonic everywhere on D. By Property (ii), $w = 0$ on bdy D, so by uniqueness of the Dirichlet problem for Laplace's equation, $w(\mathbf{x}) = 0$ for all $\mathbf{x} \in D$. Thus $G_1 = G_2$.

7.3.2. For u satisfying $\Delta u = f$, the representation formula (7.2.1) becomes

$$u(\mathbf{x}_0) = \iint_{\text{bdy } D} \left(u\frac{\partial v}{\partial n} - \frac{\partial u}{\partial n}v \right) dS + \iiint_D vf\, d\mathbf{x},$$

where $v = 1/(4\pi|\mathbf{x} - \mathbf{x}_0|)$. To see this, apply Green's Second Identity to the pair u, v on the region D_ϵ to get

$$\iiint_{D_\epsilon} -vf\, d\mathbf{x} = \iint_{\text{bdy } D_\epsilon} \left(u\frac{\partial v}{\partial n} - \frac{\partial u}{\partial n}v \right) dS. \tag{S-2}$$

The left hand side is

$$\iiint_D -vf\, d\mathbf{x} - \iiint_{B_\epsilon} -vf\, d\mathbf{x}.$$

Letting $r = |\mathbf{x} - \mathbf{x}_0|$ and using the fact that f is continuous and can be bounded by a constant C on the domain D, we have

$$\left| \iiint_{B_\epsilon} -vf\, d\mathbf{x} \right| \leq \frac{C}{4\pi} \int_0^{2\pi} \int_0^{\pi} \int_0^\epsilon r\sin\theta\, dr\, d\theta\, d\phi = \frac{C\epsilon^2}{2},$$

which goes to zero as ϵ approaches zero. Hence the left hand side of (S-2) approaches

$$\iiint_D -vf\, d\mathbf{x}.$$

The right hand side of (S-2) is exactly the expression in the proof of (7.2.1), and it is shown in the proof that is approaches

$$\iint_{\text{bdy } D} \left(u\frac{\partial v}{\partial n} - \frac{\partial u}{\partial n}v \right) dS - u(\mathbf{x}_0)$$

as ϵ goes to zero.

Now, following the proof of Theorem 7.3.1, write $G(\mathbf{x}, \mathbf{x}_0) = v(\mathbf{x}) + H(\mathbf{x})$. By Properties (i) and (iii), H is harmonic everywhere on D. Green's Second Identity applied to u and H then states

$$0 = \iiint_D Hf\, d\mathbf{x} + \iint_{\text{bdy } D} \left(u\frac{\partial H}{\partial n} - \frac{\partial u}{\partial n}H \right) dS.$$

Adding this to the representation formula above gives

$$u(\mathbf{x}_0) = \iiint_D Gf\, d\mathbf{x} + \iint_{\text{bdy } D} \left(u\frac{\partial G}{\partial n} - \frac{\partial u}{\partial n}G \right) dS,$$

and since G vanishes on bdy D, this becomes

$$u(\mathbf{x}_0) = \iiint_D G(\mathbf{x}, \mathbf{x}_0)f(\mathbf{x})\, d\mathbf{x} + \iint_{\text{bdy } D} u(\mathbf{x})\frac{\partial G}{\partial n}(\mathbf{x}, \mathbf{x}_0)\, dS.$$

78

Hence the solution of $\Delta u = f$ in D, $u = h$ on bdy D is

$$u(\mathbf{x}_0) = \iiint_D G(\mathbf{x}, \mathbf{x}_0) f(\mathbf{x}) \, d\mathbf{x} + \iint_{\text{bdy } D} h(\mathbf{x}) \frac{\partial G}{\partial n}(\mathbf{x}, \mathbf{x}_0) \, dS.$$

Section 7.4

7.4.1. Since $G''(x) = 0$ for $x \neq x_0$,

$$G(x) = \begin{cases} a_1 x + b_1 & 0 < x < x_0 \\ a_2 x + b_2 & x_0 < x < l. \end{cases}$$

Now $G(0) = 0$ implies $b_1 = 0$ and $G(l) = 0$ implies $b_2 = -a_2 l$, so that

$$G(x) = \begin{cases} a_1 x & 0 < x < x_0 \\ a_2 x - a_2 l & x_0 < x < l. \end{cases}$$

Next, $G(x)$ is continuous at x_0 (since $G(x) + \frac{1}{2}|x - x_0|$ is continuous at x_0) and thus $a_1 x_0 = a_2 x_0 - a_2 l$, which implies $a_1 = a_2 \left(1 - \frac{l}{x_0}\right)$. Lastly, the fact that $G(x) + \frac{1}{2}|x - x_0|$ is harmonic at x_0 means that the derivative of $G(x) + \frac{1}{2}|x - x_0|$ to be continuous at x_0.

$$x < x_0 \implies G(x) + \frac{1}{2}|x - x_0| = a_1 x + \frac{1}{2}(x_0 - x)$$

$$\implies \frac{d}{dx}\left(G(x) + \frac{1}{2}|x - x_0|\right) = a_1 - \frac{1}{2}$$

$$x > x_0 \implies G(x) + \frac{1}{2}|x - x_0| = a_2 x - a_2 l + \frac{1}{2}(x - x_0)$$

$$\implies \frac{d}{dx}\left(G(x) + \frac{1}{2}|x - x_0|\right) = a_2 + \frac{1}{2}.$$

Therefore, we need $a_2 + \frac{1}{2} = a_1 - \frac{1}{2}$. Using the fact that $a_1 = a_2\left(1 - \frac{\ell}{x_0}\right)$, we see that

$$a_2 + \frac{1}{2} = a_2\left(1 - \frac{\ell}{x_0}\right) - \frac{1}{2},$$

which implies $a_2 = -x_0/l$ and $a_1 = -x_0/l + 1$. Therefore, we conclude that

$$G(x) = \begin{cases} \left(-\dfrac{x_0}{l} + 1\right) x & 0 < x < x_0 \\ -\dfrac{x_0}{l}(x - l) & x_0 < x < l. \end{cases}$$

7.4.3. We first claim that

$$\frac{z_0}{2\pi} \iint \frac{1}{[(x - x_0)^2 + (y - y_0)^2 + z_0^2]^{3/2}} \, dx \, dy = 1.$$

79

To prove this we make the change of variables $\tilde{x} = (x - x_0)$ and $\tilde{y} = (y - y_0)$. Therefore,

$$\frac{z_0}{2\pi} \iint \frac{1}{[(x-x_0)^2 + (y-y_0)^2 + z_0^2]^{3/2}} \, dx \, dy = \frac{z_0}{2\pi} \iint \frac{1}{[\tilde{x}^2 + \tilde{y}^2 + z_0^2]^{3/2}} \, d\tilde{x} \, d\tilde{y}.$$

Now we write this in polar coordinates,

$$\frac{z_0}{2\pi} \int_0^{2\pi} \int_0^\infty \frac{1}{[r^2 + z_0^2]^{3/2}} r \, dr \, d\theta = z_0 \int_0^\infty \frac{1}{(s^2+1)^{3/2} z_0^3} z_0^2 s \, ds = 1.$$

where we have substituted $r = z_0 s$.

Therefore,

$$h(x_0, y_0) = \frac{z_0}{2\pi} \iint \frac{h(x_0, y_0)}{[(x-x_0)^2 + (y-y_0)^2 + z_0^2]^{3/2}} \, dx \, dy.$$

so that by equation (7.4.3),

$$|u(x_0, y_0, z_0) - h(x_0, y_0)| \leq \frac{z_0}{2\pi} \iint \frac{|h(x, y) - h(x_0, y_0)|}{[(x-x_0)^2 + (y-y_0)^2 + z_0^2]^{3/2}} \, dx \, dy.$$

Solution 1. Given $\epsilon > 0$ the continuity of h implies that there is some $\delta > 0$ such that $\sup_{B_\delta} |h(x, y) - h(x_0, y_0)| < \epsilon/2$, where B_δ is the ball of radius δ centered at (x_0, y_0). Next, assuming that $|h(x, y)|$ is bounded, say by M, we have $|h(x, y) - h(x_0, y_0)| \leq 2M$ for all (x, y). So we have

$$|u(x_0, y_0, z_0) - h(x_0, y_0)| = \frac{z_0}{2\pi} \iint_{B_\delta} \frac{|h(x, y) - h(x_0, y_0)|}{[(x-x_0)^2 + (y-y_0)^2 + z_0^2]^{3/2}} \, dx \, dy$$

$$+ \frac{z_0}{2\pi} \iint_{\mathbb{R}^2 \backslash B_\delta} \frac{|h(x, y) - h(x_0, y_0)|}{[(x-x_0)^2 + (y-y_0)^2 + z_0^2]^{3/2}} \, dx \, dy$$

$$\equiv A + B,$$

where

$$A \leq \sup_{B_\delta} |h(x, y) - h(x_0, y_0)| < \epsilon/2$$

and

$$B \leq \frac{z_0}{2\pi} \iint_{\mathbb{R}^2 \backslash B_\delta} \frac{2M}{[(x-x_0)^2 + (y-y_0)^2 + z_0^2]^{3/2}} \, dx \, dy$$

$$= \frac{z_0}{\sqrt{\delta^2 + z_0^2}} < \frac{z_0}{\delta} < \epsilon/2,$$

provided $0 < z_0 < \delta\epsilon/2$.

Solution 2. Let $x = x_0 + z_0 s \cos\theta$ and $y = y_0 + z_0 s \sin\theta$. Then

$$\frac{z_0}{2\pi} \iint \frac{|h(x, y) - h(x_0, y_0)|}{[(x-x_0)^2 + (y-y_0)^2 + z_0^2]^{3/2}} \, dx \, dy =$$

$$\int_0^{2\pi} \int_0^\infty |h(x_0 + z_0 s \cos\theta, y_0 + z_0 s \sin\theta) - h(x_0, y_0)| \cdot \frac{s \, ds}{(s^2+1)^{3/2}} \frac{d\theta}{2\pi}.$$

80

We split up the s integral into $\int_0^R + \int_R^\infty$. Since h is bounded, by say M, we have

$$\int_0^{2\pi} \int_R^\infty \cdots \leq 2M \int_R^\infty \frac{s\,ds}{(s^2+1)^{3/2}} = \frac{2M}{(R^2+1)^{1/2}} < \epsilon/2,$$

provided R is chosen sufficiently large. For this choice of R, continuity of h implies that the term

$$|h(x_0 + z_0 s \cos\theta, y_0 + z_0 s \sin\theta) - h(x_0, y_0)|$$

will be less than $\epsilon/2$ for all $s \in [0, R]$ provided z_0 is sufficiently small. With this choice of z_0 we have

$$\int_0^{2\pi} \int_0^R \cdots \leq \left(\frac{\epsilon}{2}\right) \int_0^R \frac{s\,ds}{(s^2+1)^{3/2}} < \epsilon/2.$$

Thus $|u(x_0, y_0, z_0) - h(x_0, y_0)| < \epsilon$ provided z_0 is sufficiently small.

7.4.6.

(a) Given $\mathbf{x}_0 = (x_0, y_0)$ in the half-plane D, let $\mathbf{x}_0^* = (x_0, -y_0)$ be its reflection across the boundary. Then let

$$G(\mathbf{x}, \mathbf{x}_0) = \Phi(\mathbf{x} - \mathbf{x}_0) - \Phi(\mathbf{x} - \mathbf{x}_0^*) = \frac{1}{2\pi} \left(\ln|\mathbf{x} - \mathbf{x}_0| - \ln|\mathbf{x} - \mathbf{x}_0^*| \right).$$

Since $\mathbf{x}_0^*$ is not in the half-plane, $\Phi(\mathbf{x} - \mathbf{x}_0^*)$ is harmonic on the half-plane. Also, since $|\mathbf{x} - \mathbf{x}_0| = |\mathbf{x} - \mathbf{x}_0^*|$ for all $\mathbf{x}$ in bdy D, $G(\mathbf{x}, \mathbf{x}_0)$ vanishes for $\mathbf{x}$ in bdy D. Thus $G(\mathbf{x}, \mathbf{x}_0)$ is the Green's function for D.

(b) On bdy D, $\frac{\partial}{\partial n} = -\frac{\partial}{\partial y}$. In coordinates,

$$G(\mathbf{x}, \mathbf{x}_0) = \frac{1}{4\pi} \left(\ln[(x - x_0)^2 + (y - y_0)^2] - \ln[(x - x_0)^2 + (y + y_0)^2] \right),$$

and thus for $\mathbf{x} = (x, 0)$ in bdy D, we have

$$\frac{\partial G}{\partial n}(\mathbf{x}, \mathbf{x}_0) = \frac{y_0}{\pi[(x - x_0)^2 + y_0^2]}.$$

The solution of the Dirichlet problem on the half-plane is therefore

$$u(x_0, y_0) = \frac{y_0}{\pi} \int_{-\infty}^\infty \frac{h(x)}{(x - x_0)^2 + y_0^2}\,dx.$$

(c) With $h(x) \equiv 1$ the formula from part (b) becomes

$$\frac{y_0}{\pi} \int_{-\infty}^\infty \frac{1}{(x - x_0)^2 + y_0^2}\,dx = \frac{y_0}{\pi} \int_{-\infty}^\infty \frac{1}{x^2 + y_0^2}\,dx = \left[\frac{1}{\pi} \arctan(x/y_0) \right]_{-\infty}^\infty = 1,$$

so $u(x, y) \equiv 1$.

7.4.7.

(a) If $u(x, y) = f(x/y)$, then $u_{xx} + u_{yy} = y^{-2}f''(x/y) + x^2y^{-4}f''(x/y) + 2xy^{-3}f'(x/y)$, so if u is harmonic,
$$\left(1 + (x/y)^2\right) f''(x/y) + 2(x/y)f'(x/y) = 0,$$
and therefore f satisfies $(1 + s^2)f''(s) + 2sf'(s) = 0$. This can be rewritten $((1 + s^2)f'(s))' = 0$, so $f'(s) = A(1 + s^2)^{-1}$ and $f(s) = A\arctan(s) + B$.

(b) Since $s = \frac{x}{y} = \frac{r\cos\theta}{r\sin\theta} = \cot\theta$, the solution u does not depend on r and therefore $\partial u/\partial r \equiv 0$. Alternatively,
$$\frac{\partial u}{\partial r} = \frac{\partial u}{\partial x}\frac{\partial x}{\partial r} + \frac{\partial u}{\partial y}\frac{\partial y}{\partial r} = \frac{1}{y}f'\left(\frac{x}{y}\right)\cdot\frac{x}{r} - \frac{x}{y^2}f'\left(\frac{x}{y}\right)\cdot\frac{y}{r} = 0.$$

(c) If $u_r \equiv 0$, then u depends only on θ. But θ depends only on the quotient x/y since $\theta = \operatorname{arccot}(x/y)$. Therefore u depends only on x/y.

(d) From the solution in part (a), we see that for $x > 0$,
$$\lim_{y \to 0} u(x, y) = \lim_{y \to 0} A\arctan(x/y) + B = \frac{1}{2}\pi A + B,$$
and for $x < 0$,
$$\lim_{y \to 0} u(x, y) = \lim_{y \to 0} A\arctan(x/y) + B = -\frac{1}{2}\pi A + B.$$

(e) The solution from part (a) is $u(x, y) = A\arctan(x/y) + B$. Inserting the boundary condition
$$h(x) = \begin{cases} -\frac{1}{2}\pi A + B & \text{for } x < 0 \\ \frac{1}{2}\pi A + B & \text{for } x > 0 \end{cases}$$
into the solution of the Dirichlet problem for the half plane in part (b) of Exercise 7.4.6. gives
$$u(x_0, y_0) = \frac{y_0}{\pi}\int_{-\infty}^{0} \frac{-\frac{1}{2}\pi A + B}{(x - x_0)^2 + y_0^2}\,dx + \frac{y_0}{\pi}\int_{0}^{\infty} \frac{\frac{1}{2}\pi A + B}{(x - x_0)^2 + y_0^2}\,dx$$
$$= \frac{-\frac{1}{2}\pi A + B}{\pi}\arctan\left(\frac{x - x_0}{y_0}\right)\Big|_{-\infty}^{0} + \frac{\frac{1}{2}\pi A + B}{\pi}\arctan\left(\frac{x - x_0}{y_0}\right)\Big|_{0}^{\infty}$$
$$= \frac{-\frac{1}{2}\pi A + B}{\pi}\left(\frac{\pi}{2} - \arctan(x_0/y_0)\right) + \frac{\frac{1}{2}\pi A + B}{\pi}\left(\frac{\pi}{2} + \arctan(x_0/y_0)\right)$$
$$= A\arctan(x_0/y_0) + B,$$
in agreement with the solution from part (a).

7.4.9. The vector $\mathbf{n} = (a, b, c)$ is normal to the plane $ax + by + cz = 0$. Given $\mathbf{x}_0 = (x_0, y_0, z_0)$ in the half-space D, its reflection across the plane is of the form $\mathbf{x}_0^* = \mathbf{x}_0 + k\mathbf{n}$ for some scalar k and should satisfy $|\mathbf{x} - \mathbf{x}_0| = |\mathbf{x} - \mathbf{x}_0^*|$ for every $\mathbf{x}$ in the plane $ax + by + cz = 0$. Thus we want

$$
\begin{aligned}
0 &= |\mathbf{x} - \mathbf{x}_0|^2 - |\mathbf{x} - \mathbf{x}_0^*|^2 \\
&= |\mathbf{x}|^2 - 2\mathbf{x} \cdot \mathbf{x}_0 + |\mathbf{x}_0|^2 - |\mathbf{x}|^2 + 2\mathbf{x} \cdot \mathbf{x}_0^* - |\mathbf{x}_0^*|^2 \\
&= |\mathbf{x}_0|^2 - |\mathbf{x}_0^*|^2 + 2\mathbf{x} \cdot (\mathbf{x}_0^* - \mathbf{x}_0)
\end{aligned}
$$

for all $\mathbf{x}$ in the plane. But $\mathbf{x} \cdot (\mathbf{x}_0^* - \mathbf{x}_0) = k\mathbf{x} \cdot \mathbf{n} = k(ax + by + cz) = 0$, so we just need to find k such that $|\mathbf{x}_0|^2 = |\mathbf{x}_0^*|^2$. We therefore have

$$
\mathbf{x}_0 \cdot \mathbf{x}_0 = \mathbf{x}_0 \cdot \mathbf{x}_0 + 2k\mathbf{x}_0 \cdot \mathbf{n} + k^2\mathbf{n} \cdot \mathbf{n}
$$

which implies

$$
k = -\frac{2\mathbf{x}_0 \cdot \mathbf{n}}{\mathbf{n} \cdot \mathbf{n}} = -2\frac{ax_0 + by_0 + cz_0}{a^2 + b^2 + c^2}.
$$

With $\mathbf{x}_0^*$ defined in this way, the Green's function for this tilted half-space is therefore

$$
G(\mathbf{x}, \mathbf{x}_0) = \Phi(\mathbf{x} - \mathbf{x}_0) - \Phi(\mathbf{x} - \mathbf{x}_0^*) = -\frac{1}{4\pi|\mathbf{x} - \mathbf{x}_0|} + \frac{1}{4\pi|\mathbf{x} - \mathbf{x}_0^*|}.
$$

7.4.11.

(a) In two dimensions, using the solution $(1/2\pi)\log r$ of Laplace's equation, we have the following analogue of part (iii) in the definition of the Green's function:

- $G(\mathbf{x}) - (1/2\pi)\log(|\mathbf{x} - \mathbf{x}_0|)$ is finite at $\mathbf{x}_0$ and has continuous second derivatives everywhere in D and is harmonic at $\mathbf{x}_0$.

Since $\log \rho$ and $\log\left(\frac{r_0}{a}\rho^*\right) = \log(r_0/a) + \log(\rho^*)$ are translations of $\log r$, they are harmonic, except at x_0 and x_0^* respectively, so G is harmonic within D, except at x_0, and therefore condition (i) holds. Condition (iii) holds since $G(\mathbf{x}) - (1/2\pi)\log(|\mathbf{x} - \mathbf{x}_0|) = -(1/2\pi)\log\left(\frac{r_0}{a}\rho^*\right)$ is harmonic everywhere in D. To prove (ii), use the fact that, by equation (7.4.8), $\frac{r_0}{a}\rho^* = \rho$ for $|\mathbf{x}| = a$, and therefore

$$
G(\mathbf{x}, \mathbf{x}_0) = \frac{1}{2\pi}\log \rho - \frac{1}{2\pi}\log \rho = 0
$$

for $|\mathbf{x}| = a$.

(b) To use formula (7.3.1), we need to compute $\partial G/\partial n$. First,

$$
\nabla G = \frac{1}{2\pi}\left(\frac{\mathbf{x} - \mathbf{x}_0}{|\mathbf{x} - \mathbf{x}_0|^2} - \frac{\mathbf{x} - \mathbf{x}_0^*}{|\mathbf{x} - \mathbf{x}_0^*|^2}\right) = \frac{1}{2\pi}\left(\frac{\mathbf{x} - \mathbf{x}_0}{\rho^2} - \frac{\mathbf{x} - \mathbf{x}_0^*}{(\rho^*)^2}\right)
$$

Since $\mathbf{x}_0^* = a^2\mathbf{x}_0/r_0^2$ and $\rho^* = \rho a/r_0$ for $|x| = a$, this becomes

$$
\nabla G = \frac{1}{2\pi}\left(\frac{\mathbf{x} - \mathbf{x}_0}{\rho^2} - \frac{\mathbf{x} - a^2\mathbf{x}_0/r_0^2}{a^2\rho^2/r_0^2}\right) = \frac{(a^2 - r_0^2)\mathbf{x}}{2\pi\rho^2 a^2}
$$

for $\mathbf{x}$ on the boundary of D. Hence

$$\frac{\partial G}{\partial n} = \nabla G \cdot \frac{\mathbf{x}}{a} = \frac{(a^2 - r_0^2)}{2\pi\rho^2 a}.$$

Applying formula (7.4.1) then gives

$$u(\mathbf{x}_0) = \frac{a^2 - |\mathbf{x}_0|^2}{2\pi a} \int_{|\mathbf{x}|=a} \frac{u(\mathbf{x})}{|\mathbf{x} - \mathbf{x}_0|^2} \, ds,$$

which is precisely Poisson's formula (6.3.14).

7.4.13. For $\mathbf{x}_0 = (x_0, y_0, z_0)$ in the half-ball D, define $\mathbf{x}_0^\sharp = (x_0, y_0, -z_0)$. Let G_B denote the Green's function for the entire ball. then define

$$G(\mathbf{x}, \mathbf{x}_0) = G_B(\mathbf{x}, \mathbf{x}_0) - G_B(\mathbf{x}, \mathbf{x}_0^\sharp)$$

$$= -\frac{1}{4\pi|\mathbf{x} - \mathbf{x}_0|} + \frac{1}{4\pi|\mathbf{x} - \mathbf{x}_0^*|} + \frac{1}{4\pi|\mathbf{x} - \mathbf{x}_0^\sharp|} - \frac{1}{4\pi|\mathbf{x} - (\mathbf{x}_0^\sharp)^*|}.$$

To verify that this is the Green's function for D, notice that the first term is $\Phi(\mathbf{x} - \mathbf{x}_0)$ and the other terms are the fundamental solutions centered at $\mathbf{x}_0^*, \mathbf{x}_0^\sharp$ and $(\mathbf{x}_0^\sharp)^*$, all of which are outside of D. Thus the other three terms are harmonic in D. Now the boundary of D consists of the upper hemisphere and the disk in the xy-plane. Since both $G_B(\mathbf{x}, \mathbf{x}_0)$ and $G_B(\mathbf{x}, \mathbf{x}_0^\sharp)$ vanish for $\mathbf{x}$ on the hemisphere, so does $G(\mathbf{x}, \mathbf{x}_0)$. Also, $|\mathbf{x} - \mathbf{x}_0^\sharp| = |\mathbf{x} - \mathbf{x}_0|$ for $\mathbf{x}$ in the disk, so the first and third terms cancel for $\mathbf{x}$ in the disk. Likewise, since $(\mathbf{x}_0^\sharp)^* = (\mathbf{x}_0^*)^\sharp$, we have $|\mathbf{x} - (\mathbf{x}_0^\sharp)^*| = |\mathbf{x} - (\mathbf{x}_0^*)^\sharp| = |\mathbf{x} - \mathbf{x}_0^*|$ for points $\mathbf{x}$ in the disk. Thus the second and fourth terms cancel for $\mathbf{x}$ in the disk. So $G(\mathbf{x}, \mathbf{x}_0) = 0$ for all $\mathbf{x}$ in bdy D, and therefore G is the Green's function for D.

7.4.15.

(a) For $u(x, y) = v(x^2 - y^2, 2xy)$, we have

$$u_x = 2xv_x + 2yv_y$$
$$u_{xx} = 4x^2v_{xx} + 8xyv_{xy} + 4y^2v_{yy} + 2v_x$$
$$u_y = -2yv_x + 2xv_y$$
$$u_{yy} = 4y^2v_{xx} - 8xyv_{xy} + 4x^2v_{yy} - 2v_x,$$

and therefore

$$u_{xx} + u_{yy} = (4x^2 + 4y^2)(v_{xx} + v_{yy}) = 0$$

by the assumption that v is harmonic.

(b) In polar coordinates,

$$(x^2 - y^2, 2xy) = (r^2(\cos^2\theta - \sin^2\theta), 2r^2\cos\theta\sin\theta) = (r^2\cos(2\theta), r^2\sin(2\theta)).$$

In the first quadrant, r goes from 0 to ∞ and θ ranges from 0 to $\pi/2$, so r^2 goes from 0 to ∞ and 2θ goes from 0 to π. Therefore the image of the first quadrant is the upper half plane.

7.4.17.

(a) For $\mathbf{x}_0 = (x_0, y_0)$, define $\mathbf{x}_0^* = (x_0, -y_0)$ and $\mathbf{x}_0' = (-x_0, y_0)$. Recall that

$$G_H(\mathbf{x}, \mathbf{x}_0) = \Phi(\mathbf{x} - \mathbf{x}_0) - \Phi(\mathbf{x} - \mathbf{x}_0^*) = \frac{1}{2\pi}\left(\ln|\mathbf{x} - \mathbf{x}_0| - \ln|\mathbf{x} - \mathbf{x}_0^*|\right)$$

is the Green's function for the upper half-plane. Define

$$G(\mathbf{x}, \mathbf{x}_0) = G_H(\mathbf{x}, \mathbf{x}_0) - G_H(\mathbf{x}, \mathbf{x}_0')$$

$$= \frac{1}{2\pi}\left(\ln|\mathbf{x} - \mathbf{x}_0| - \ln|\mathbf{x} - \mathbf{x}_0^*| - \ln|\mathbf{x} - \mathbf{x}_0'| + \ln|\mathbf{x} - (\mathbf{x}_0')^*|\right).$$

The first term is the fundamental solution centered at $\mathbf{x}_0$, and the other three terms are harmonic in Q, being fundamental solutions centered at points outside of Q. Since $G_H(\mathbf{x}, \mathbf{x}_0)$ and $G_H(\mathbf{x}, \mathbf{x}_0')$ both vanish on the x-axis, so does $G(\mathbf{x}, \mathbf{x}_0)$. Since $|\mathbf{x} - \mathbf{x}_0'| = |\mathbf{x} - \mathbf{x}_0|$ for $\mathbf{x}$ on the y-axis, the first and third terms cancel for $\mathbf{x}$ on the y-axis. Likewise, since $(\mathbf{x}_0')^* = (\mathbf{x}_0^*)'$, the second and fourth terms cancel for $\mathbf{x}$ on the y-axis. Hence $G(\mathbf{x}, \mathbf{x}_0) = 0$ for all $\mathbf{x} \in$ bdy Q, so G is the Green's function for Q.

(b) In coordinates (x, y) and (x_0, y_0), the Green's function takes the form

$$G(x, y, x_0, y_0) = \frac{1}{4\pi}\ln\left[(x - x_0)^2 + (y - y_0)^2\right] - \frac{1}{4\pi}\ln\left[(x - x_0)^2 + (y + y_0)^2\right]$$

$$- \frac{1}{4\pi}\ln\left[(x + x_0)^2 + (y - y_0)^2\right] + \frac{1}{4\pi}\ln\left[(x + x_0)^2 + (y + y_0)^2\right].$$

On the bottom side, $y = 0$ and $\frac{\partial G}{\partial n} = -\frac{\partial G}{\partial y}$, so for $\mathbf{x} = (x, 0)$ on the bottom side we have

$$\frac{\partial G}{\partial n}(\mathbf{x}, \mathbf{x}_0) = \frac{y_0}{\pi}\left(\frac{1}{(x - x_0)^2 + y_0^2} - \frac{1}{(x + x_0)^2 + y_0^2}\right).$$

Similarly, for $\mathbf{x} = (0, y)$ on the left side, we have

$$\frac{\partial G}{\partial n}(\mathbf{x}, \mathbf{x}_0) = \frac{x_0}{\pi}\left(\frac{1}{x_0^2 + (y - y_0)^2} - \frac{1}{x_0^2 + (y + y_0)^2}\right).$$

Thus the solution of the Dirichlet problem is

$$u(x_0, y_0) = \frac{y_0}{\pi}\int_0^\infty h(x)\left(\frac{1}{(x - x_0)^2 + y_0^2} - \frac{1}{(x + x_0)^2 + y_0^2}\right)dx$$

$$+ \frac{x_0}{\pi}\int_0^\infty g(y)\left(\frac{1}{x_0^2 + (y - y_0)^2} - \frac{1}{x_0^2 + (y + y_0)^2}\right)dy.$$

7.4.21. First, recall that Theorem 7.2.1 states that for any function u that is harmonic in D,

$$u(\mathbf{x}_0) = \iint_{\text{bdy } D} u(\mathbf{x})\frac{\partial \Phi}{\partial n}(\mathbf{x} - \mathbf{x}_0) - \Phi(\mathbf{x} - \mathbf{x}_0)\frac{\partial u}{\partial n}(\mathbf{x})\,dS,$$

where $\Phi = -\frac{1}{4\pi r}$ is the fundamental solution of Laplace's equation. By Green's second identity, we can add any harmonic function H to Φ and the identity still holds. We now define a Neumann function for the domain D to be a function of the form $N(\mathbf{x}, \mathbf{x}_0) = \Phi(\mathbf{x} - \mathbf{x}_0) + H(\mathbf{x}, \mathbf{x}_0)$, where H is harmonic in $\mathbf{x}$ on all of D, and

$$\frac{\partial N}{\partial n}(\mathbf{x}, \mathbf{x}_0) = C$$

for some constant C. Then

$$u(\mathbf{x}_0) = \iint_{\text{bdy } D} C \cdot u(\mathbf{x}) - N(\mathbf{x}, \mathbf{x}_0)\frac{\partial u}{\partial n}(\mathbf{x}) \, dS.$$

Notice that if we apply this result to the constant function $u \equiv 1$, we get

$$1 = \iint_{\text{bdy } D} C \, dS = C \cdot \text{area(bdy } D)$$

so there is only one possible value for C, determined by the surface area of bdy D. So if u is a harmonic function on D such that

$$\iint_{\text{bdy } D} u(\mathbf{x}) \, dS = 0, \tag{S-3}$$

then

$$u(\mathbf{x}_0) = -\iint_{\text{bdy } D} N(\mathbf{x}, \mathbf{x}_0)\frac{\partial u}{\partial n}(\mathbf{x}) \, dS. \tag{S-4}$$

Turning now to the Neumann problem

$$\begin{cases} \Delta u = 0 & \text{in } D \\ \frac{\partial u}{\partial n} = h & \text{in bdy } D \end{cases}$$

for Laplace's equation, first recall that solutions are unique up to addition of a constant. Thus, if u is a solution, we can add an appropriate constant K to u in order to make its integral over bdy D vanish (in fact $K = -(\iint_{\text{bdy } D} u \, dS)/\text{area(bdy } D)$). Hence we can assume without loss of generality that (S-3) holds for our solution u. Thus our solution satisfies (S-4) and we have the following theorem.

Theorem. The solution of the Neumann problem

$$\begin{cases} \Delta u = 0 & \text{in } D \\ \frac{\partial u}{\partial n} = h & \text{in bdy } D \end{cases}$$

is

$$u(\mathbf{x}_0) = C - \iint_{\text{bdy } D} N(\mathbf{x}, \mathbf{x}_0)h(\mathbf{x}) \, dS$$

for any constant C.

7.4.22. The function $v = u_y$ satisfies Laplace's equation with boundary condition $v(x, 0) = h(x)$, so by the solution of Exercise 7.4.6(b),

$$v(x_0, y_0) = \frac{y_0}{\pi} \int_{-\infty}^{\infty} \frac{h(x)}{(x - x_0)^2 + y_0^2} \, dx = \frac{1}{2\pi}\frac{\partial}{\partial y_0} \int_{-\infty}^{\infty} h(x) \log\left((x - x_0)^2 + y_0^2\right) \, dx.$$

86

Therefore

$$u(x_0, y_0) = \frac{1}{2\pi} \int_{-\infty}^{\infty} h(x) \log\left((x - x_0)^2 + y_0^2\right) dx + f(y_0).$$

where $f(y_0)$ is any harmonic function such that $f'(0) = 0$. The only such function is $f \equiv C$ for some constant C.

7.4.24. The boundary of the half-space D has infinite surface area, so by the argument given in the solution of Exercise 7.4.21, a Neumann function should satisfy $\frac{\partial N}{\partial n} = 0$ on bdy D. Since $\frac{\partial}{\partial n} = -\frac{\partial}{\partial z}$ on bdy D, we have

$$\frac{\partial \Phi}{\partial n}(\mathbf{x}, \mathbf{x}_0) = -\frac{\partial \Phi}{\partial z}(\mathbf{x}, \mathbf{x}_0) = \frac{z_0}{4\pi|\mathbf{x} - \mathbf{x}_0|^3}.$$

Thus, for $\mathbf{x}_0 = (x_0, y_0, z_0)$ in D, if we define $\mathbf{x}_0^* = (x_0, y_0, -z_0)$ then we should *add* the corresponding fundamental solutions to obtain a zero normal derivative. Thus

$$N(\mathbf{x}, \mathbf{x}_0) = \Phi(\mathbf{x} - \mathbf{x}_0) + \Phi(\mathbf{x} - \mathbf{x}_0^*) = -\frac{1}{4\pi|\mathbf{x} - \mathbf{x}_0|} - \frac{1}{4\pi|\mathbf{x} - \mathbf{x}_0^*|}$$

is a Neumann function for D. Now for $\mathbf{x}$ in bdy D, $|\mathbf{x} - \mathbf{x}_0^*| = |\mathbf{x} - \mathbf{x}_0|$, so we have

$$N(\mathbf{x}, \mathbf{x}_0) = -\frac{1}{2\pi|\mathbf{x} - \mathbf{x}_0|}$$

for $\mathbf{x} \in$ bdy D. Applying the result from Exercise 7.4.21 gives

$$u(\mathbf{x}_0) = C + \iint_{\text{bdy } D} \frac{h(\mathbf{x})}{2\pi|\mathbf{x} - \mathbf{x}_0|} \, dS$$

$$= C + \int_{-\infty}^{\infty} \int_{-\infty}^{\infty} \frac{h(x, y)}{2\pi((x - x_0)^2 + (y - y_0)^2 + z_0^2)^{1/2}} \, dy \, dx.$$

Chapter 8

Section 8.1

8.1.1.

(a) If u is a C^3 function then

$$u(x + \Delta x) = u(x) + u'(x)\Delta x + \frac{1}{2}u''(x)(\Delta x)^2 + \frac{1}{6}u'''(x)(\Delta x)^3 + o((\Delta x)^3)$$

$$u(x - \Delta x) = u(x) - u'(x)\Delta x + \frac{1}{2}u''(x)(\Delta x)^2 - \frac{1}{6}u'''(x)(\Delta x)^3 + o((\Delta x)^3),$$

so

$$u'(x) = \frac{u(x + \Delta x) - u(x - \Delta x)}{2\Delta x} - \frac{1}{6}u'''(x)(\Delta x)^2 + \frac{o((\Delta x)^3)}{2\Delta x}.$$

Now the last term is $o((\Delta x)^2)$, and the second to last term is $O((\Delta x)^2)$, and since $o((\Delta x)^2)$ implies $O((\Delta x)^2)$ we have

$$u'(x) = \frac{u(x + \Delta x) - u(x - \Delta x)}{2\Delta x} + O((\Delta x)^2).$$

(b) If u is a C^2 function then

$$u(x + \Delta x) = u(x) + u'(x)\Delta x + \frac{1}{2}u''(x)(\Delta x)^2 + o((\Delta x)^2)$$

$$u(x - \Delta x) = u(x) - u'(x)\Delta x + \frac{1}{2}u''(x)(\Delta x)^2 + o((\Delta x)^2),$$

so

$$u'(x) = \frac{u(x + \Delta x) - u(x - \Delta x)}{2\Delta x} + \frac{o((\Delta x)^2)}{2\Delta x}.$$

Hence the error is $o(\Delta x)$, which implies it is $O(\Delta x)$.

8.1.3. If u is C^5, then

$$u(x + \Delta x) = u(x) + u'(x)\Delta x + \frac{1}{2}u''(x)(\Delta x)^2 + \frac{1}{6}u'''(x)(\Delta x)^3 + \frac{1}{24}u''''(x)(\Delta x)^4 + O((\Delta x)^5)$$

$$u(x - \Delta x) = u(x) - u'(x)\Delta x + \frac{1}{2}u''(x)(\Delta x)^2 - \frac{1}{6}u'''(x)(\Delta x)^3 + \frac{1}{24}u''''(x)(\Delta x)^4 + O((\Delta x)^5)$$

$$u(x + 2\Delta x) = u(x) + 2u'(x)\Delta x + 2u''(x)(\Delta x)^2 + \frac{4}{3}u'''(x)(\Delta x)^3 + \frac{2}{3}u''''(x)(\Delta x)^4 + O((\Delta x)^5)$$

$$u(x - 2\Delta x) = u(x) - 2u'(x)\Delta x + 2u''(x)(\Delta x)^2 - \frac{4}{3}u'''(x)(\Delta x)^3 + \frac{2}{3}u''''(x)(\Delta x)^4 + O((\Delta x)^5).$$

We now want to add multiples of these equations to cancel all but the $u'(x)\Delta x$ terms and error terms on the right side. Multiplying the first equation by A, the second by $-A$, the third by B and the fourth by $-B$ cancels the u, u'' and u'''' terms, and the right hand side becomes

$$(2A + 4B)u'(x)\Delta x + \left(\frac{1}{3}A + \frac{8}{3}B\right)u'''(x)(\Delta x)^3 + O((\Delta x)^5).$$

To cancel the u''' term and make the u' coefficient 1, we choose $A = 2/3$ and $B = -1/12$. Solving for u' then yields

$$u'(x) = \frac{\frac{2}{3}u(x+\Delta x) - \frac{2}{3}u(x-\Delta x) - \frac{1}{12}u(x+2\Delta x) + \frac{1}{12}u(x-2\Delta x)}{\Delta x} + O((\Delta x)^4).$$

Section 8.2

8.2.1.

(a) Here $s = 1/4$ and the scheme takes the form $u_j^{n+1} = \frac{1}{4}(u_{j+1}^n + u_{j-1}^n) + \frac{1}{2}u_j^n$ for $1 \leq j \leq 3$. The boundary conditions imply $u_0^n = u_4^n = 0$ for all n. The solution is:

$$n = 4: \quad 0 \quad \frac{99}{64} \quad \frac{35}{16} \quad \frac{99}{64} \quad 0$$

$$n = 3: \quad 0 \quad \frac{29}{16} \quad \frac{41}{16} \quad \frac{29}{16} \quad 0$$

$$n = 2: \quad 0 \quad \frac{17}{8} \quad 3 \quad \frac{17}{8} \quad 0$$

$$n = 1: \quad 0 \quad \frac{5}{2} \quad \frac{7}{2} \quad \frac{5}{2} \quad 0$$

$$n = 0: \quad 0 \quad 3 \quad 4 \quad 3 \quad 0$$

(b) Here $s = 1/4$ again so we get the same scheme as in part (a) for $1 \leq j \leq 7$, and $u_0^n = u_8^n = 0$. The solution is:

$$n = 4: \quad 0 \quad \frac{709}{512} \quad \frac{323}{128} \quad \frac{1665}{512} \quad \frac{7}{2} \quad \frac{1665}{512} \quad \frac{323}{128} \quad \frac{709}{512} \quad 0$$

$$n = 3: \quad 0 \quad \frac{93}{64} \quad \frac{337}{128} \quad \frac{27}{8} \quad \frac{29}{8} \quad \frac{27}{8} \quad \frac{337}{128} \quad \frac{93}{64} \quad 0$$

$$n = 2: \quad 0 \quad \frac{49}{32} \quad \frac{11}{4} \quad \frac{7}{2} \quad \frac{15}{4} \quad \frac{7}{2} \quad \frac{11}{4} \quad \frac{49}{32} \quad 0$$

$$n = 1: \quad 0 \quad \frac{13}{8} \quad \frac{23}{8} \quad \frac{29}{8} \quad \frac{31}{8} \quad \frac{29}{8} \quad \frac{23}{8} \quad \frac{13}{8} \quad 0$$

$$n = 0: \quad 0 \quad \frac{7}{4} \quad 3 \quad \frac{15}{4} \quad 4 \quad \frac{15}{4} \quad 3 \quad \frac{7}{4} \quad 0$$

(c) At $x = 2$, $t = 0.25$ we get $7/2$ in part(a) and $15/4$ in part (b).

89

8.2.3.

(a) Since $s = 1/2$, the scheme is $u_j^{n+1} = \frac{1}{2}(u_{j-1}^n + u_{j+1}^n)$ for $1 \leq j \leq 4$. The boundary conditions imply $u_0^n = 0$ and $u_5^n = 1$ for all n. The solution is:

$$n = 6: \quad 0 \quad \frac{7}{64} \quad \frac{7}{32} \quad \frac{29}{64} \quad \frac{11}{16} \quad 1$$

$$n = 5: \quad 0 \quad \frac{1}{16} \quad \frac{7}{32} \quad \frac{3}{8} \quad \frac{11}{16} \quad 1$$

$$n = 4: \quad 0 \quad \frac{1}{16} \quad \frac{1}{8} \quad \frac{3}{8} \quad \frac{5}{8} \quad 1$$

$$n = 3: \quad 0 \quad 0 \quad \frac{1}{8} \quad \frac{1}{4} \quad \frac{5}{8} \quad 1$$

$$n = 2: \quad 0 \quad 0 \quad 0 \quad \frac{1}{4} \quad \frac{1}{2} \quad 1$$

$$n = 1: \quad 0 \quad 0 \quad 0 \quad 0 \quad \frac{1}{2} \quad 1$$

$$n = 0: \quad 0 \quad 0 \quad 0 \quad 0 \quad 0 \quad 1$$

So $u(3, 3)$ is approximated by $29/64$.

(b) If we let $v(x, t) = u(x, t) - x/5$, then v satisfies $v_t = v_{xx}$ with $v(0, t) = v(5, t) = 0$ and $v(x, 0) = -x/5$. The Fourier sine series for $-x/5$ on the interval $(0, 5)$ has coefficients $2(-1)^n/n\pi$, so the solution of the PDE is

$$v(x, t) = \sum_{n=1}^{\infty} \frac{2(-1)^n}{n\pi} \sin \frac{n\pi x}{5} e^{-n^2\pi^2 t/25},$$

and thus

$$v(3, 3) = \sum_{n=1}^{\infty} \frac{2(-1)^n}{n\pi} \sin \frac{3n\pi}{5} e^{-3n^2\pi^2/25} \approx -0.187.$$

This implies that $u(3, 3) \approx 0.413$. The approximation in part (a) is about 0.453, so they differ by about 10%.

8.2.5. Since $s = 1/2$, the scheme is given by $u_j^{n+1} = \frac{1}{2}(u_{j-1}^n + u_{j+1}^n)$ for $1 \leq j \leq 5$. The boundary condition $u(0, t) = 0$ implies $u_0^n = 0$ for all n, and the condition $u_x(5, t) = 0$ implies the ghost points satisfy $u_6^n = u_4^n$ for all n. The solution is shown below. The values of the ghost points are shown in parentheses.

$$n = 6: \quad 0 \quad \frac{35}{8} \quad \frac{495}{64} \quad 10 \quad \frac{735}{64} \quad \frac{45}{4} \quad \left(\frac{735}{64}\right)$$

$$n = 5: \quad 0 \quad \frac{75}{16} \quad \frac{35}{4} \quad \frac{345}{32} \quad \frac{45}{4} \quad \frac{195}{16} \quad \left(\frac{45}{4}\right)$$

$$n = 4: \quad 0 \quad 6 \quad \frac{75}{8} \quad \frac{23}{2} \quad \frac{195}{16} \quad 11 \quad \left(\frac{195}{16}\right)$$

$$n = 3: \quad 0 \quad \frac{51}{8} \quad 12 \quad \frac{99}{8} \quad 11 \quad 12 \quad (11)$$

$$n = 2: \quad 0 \quad 10 \quad \frac{51}{4} \quad 14 \quad 12 \quad 8 \quad (12)$$

$$n = 1: \quad 0 \quad \frac{21}{2} \quad 20 \quad 15 \quad 8 \quad 9 \quad (8)$$

$$n = 0: \quad 0 \quad 24 \quad 21 \quad 16 \quad 9 \quad 0 \quad (9)$$

The approximation for $u(3,3)$ is 10.

8.2.8.

(a) The scheme is

$$u_j^{n+1} = u_j^n + \frac{s}{2}(u_{j+1}^n - 2u_j^n + u_{j-1}^n + u_{j+1}^{n+1} - 2u_j^{n+1} + u_{j-1}^{n+1}),$$

or equivalently

$$\frac{s}{2}u_{j+1}^{n+1} - (s+1)u_j^{n+1} + \frac{s}{2}u_{j-1}^{n+1} = -\frac{s}{2}u_{j+1}^n + (s-1)u_j^n - \frac{s}{2}u_{j-1}^n.$$

(b) Let $v(x,t) = u(1-x,t)$. Then $v_t(x,t) = u_t(1-x,t)$, $v_x(x,t) = -u_x(1-x,t)$ and $v_{xx}(x,t) = u_{xx}(1-x,t) = u_t(1-x,t) = v_t(x,t)$. Thus v is a solution of the heat equation. Its initial data is $v(x,0) = u(1-x,0) = \phi(1-x) = \phi(x)$. By uniqueness, it therefore follows that $v(x,t) = u(x,t)$. That is, $u(1-x,t) = u(x,t)$.

(c) First, $s = 6$. So the template is:

$$\begin{array}{ccc} 3/7 & * & 3/7 \\ 3/7 & -5/7 & 3/7 \end{array}$$

Next, the boundary condition implies $u_0^1 = u_6^1 = 0$. The scheme in part (a) then implies

$$\begin{array}{rcl}
3u_2^1 - 7u_1^1 & = & 0 \\
3u_3^1 - 7u_2^1 + 3u_1^1 & = & -3 \\
3u_4^1 - 7u_3^1 + 3u_2^1 & = & 5 \\
3u_5^1 - 7u_4^1 + 3u_3^1 & = & -3 \\
- 7u_5^1 + 3u_4^1 & = & 0.
\end{array}$$

Solving this system of simultaneous linear equations gives

$$0 \quad \frac{9}{77} \quad \frac{3}{11} \quad -\frac{37}{77} \quad \frac{3}{11} \quad \frac{9}{77} \quad 0$$

after one time step.

8.2.9. The scheme (8.2.15) can be rewritten as

$$u_j^{n+1} - u_j^n = s(1-\theta)(u_{j+1}^n - 2u_j^n + u_{j-1}^n) + s\theta(u_{j+1}^{n+1} - 2u_j^{n+1} + u_{j-1}^{n+1}).$$

Inserting $u_j^n = X_j T_n$, where $X_j = (e^{ik\Delta x})^j$ and $T_n = (\xi(k))^n$ yields

$$X_j(T_{n+1} - T_n) = (X_{j+1} - 2X_j + X_{j-1})\left[s(1-\theta)T_n + s\theta T_{n+1}\right].$$

Dividing by $u = X_j T_n$ gives

$$T_1 - 1 = (X_1 - 2 + X_{-1})\left[s(1-\theta) + s\theta T_1\right],$$

so solving for $T_1 = \xi(k)$ gives

$$\xi(k) = \frac{1 + s(1-\theta)(X_1 - 2 + X_{-1})}{1 - s\theta(X_1 - 2 + X_{-1})}.$$

It then follows from

$$X_1 - 2 + X_{-1} = e^{ik\Delta x} - 2 + e^{-ik\Delta x} = -2 + 2\cos k\Delta x$$

that

$$\xi(k) = \frac{1 - 2s(1-\theta)(1 - \cos k\Delta x)}{1 + 2s\theta(1 - \cos k\Delta x)}.$$

Since s and θ are positive, and $1 - \cos k\Delta x \geq 0$, the denominator is positive and the numerator is less than or equal to the denominator. Hence $\xi(k) \leq 1$. On the other hand, if we want $\xi(k) \geq -1$, then we need

$$1 - 2s(1-\theta)(1 - \cos k\Delta x) \geq -(1 + 2s\theta(1 - \cos k\Delta x)).$$

Rearranging terms gives

$$2s(1 - 2\theta)(1 - \cos k\Delta x) \leq 2.$$

Since s and $1 - \cos k\Delta x$ are nonnegative, this inequality holds whenever $1 - 2\theta \leq 0$. This proves (8.2.16).

Now suppose $\theta < \frac{1}{2}$. Then since $1 - \cos k\Delta x \leq 2$, the inequality above certainly holds whenever $2s(1 - 2\theta)2 \leq 2$. This proves (8.2.17).

8.2.12.

(a) The scheme is $u_j^{n+1} = \frac{1}{4}(u_{j+1}^n + u_{j-1}^n) + \frac{1}{2}u_j^n + (u_j^n)^3$. The solution is:

$n = 3$:	0.0156	0.1604	1.4730	79.1484	1.4730	0.1604	0.0156
$n = 2$:	0	0.0625	0.5156	4.25	0.5156	0.0625	0
$n = 1$:	0	0	0.25	1.5	0.25	0	0
$n = 0$:	0	0	0	1	0	0	0

Thus $u_0^3 = 79.1484$.

(b) Without the nonlinear term, the scheme becomes $u_j^{n+1} = \frac{1}{4}(u_{j+1}^n + u_{j-1}^n) + \frac{1}{2}u_j^n$. The solution in this case is:

$$n = 3: \quad 0.0156 \quad 0.0938 \quad 0.2344 \quad 0.3125 \quad 0.2344 \quad 0.0938 \quad 0.0156$$

$$n = 2: \quad\quad 0 \quad 0.0625 \quad\quad 0.25 \quad 0.375 \quad\quad 0.25 \quad 0.0625 \quad\quad 0$$

$$n = 1: \quad\quad 0 \quad\quad 0 \quad 0.25 \quad 0.5 \quad 0.25 \quad\quad 0 \quad\quad 0$$

$$n = 0: \quad\quad 0 \quad\quad 0 \quad\quad 0 \quad 1 \quad\quad 0 \quad\quad 0 \quad\quad 0$$

Thus $u_0^3 = 0.3125$.

(c) The ODE $dv/dt = v^3$ is separable.

$$\int \frac{dv}{v^3} = \int 1\, dt = t + C \quad\Longrightarrow\quad -\frac{1}{2v^2} = t + C$$

The initial condition implies $C = -1/2$, so solving gives

$$v(t) = \frac{1}{\sqrt{1 - 2t}}.$$

Thus

$$\lim_{t \to \frac{1}{2}} v(t) = +\infty$$

so the solution v "blows up" in finite time. This explains the rapid growth of the solution of the PDE at $x = 0$.

(d) The scheme is $u_j^{n+1} = \frac{1}{4}(u_{j+1}^n + u_{j-1}^n) + \frac{1}{2}u_j^n - (u_j^n)^3$, and the solution is:

$$n = 3: \quad 0.0156 \quad 0.0271 \quad\quad 0.0078 \quad -0.0078 \quad\quad 0.0078 \quad 0.0271 \quad 0.0156$$

$$n = 2: \quad\quad 0 \quad 0.0625 \quad -0.0156 \quad\quad\quad 0 \quad -0.0156 \quad 0.0625 \quad\quad 0$$

$$n = 1: \quad\quad 0 \quad\quad 0 \quad\quad 0.25 \quad -0.5 \quad\quad 0.25 \quad\quad 0 \quad\quad 0$$

$$n = 0: \quad\quad 0 \quad\quad 0 \quad\quad\quad 0 \quad 1 \quad\quad\quad 0 \quad\quad 0 \quad\quad 0$$

The ODE $dv/dt = -v^3$, $v(0) = 1$ has solution $v(t) = 1/\sqrt{2t + 1}$, which goes to zero as t goes to infinity. This explains why the solution u decays at $x = 0$ rather than blowing up as in part (a).

8.2.13.

(a) $u_j^{n+1} = \dfrac{2s(u_{j+1}^n + u_{j-1}^n) + (1 - 2s)u_j^{n-1}}{1 + 2s}$

(b) Let $u_j^n = T_n X_j$, where $T_n = (\xi(k))^n$ and $X_j = (e^{ik\Delta x})^j$. Then

$$X_j(T_{n+1} - T_{n-1}) = 2s[T_n(X_{j+1} + X_{j-1}) - X_j(T_{n+1} + T_{n-1})].$$

Dividing by $X_j T_n$ gives

$$T_1 - T_{-1} = 2s[X_1 + X_{-1} - T_1 - T_{-1}]$$

so

$$\xi^2(1 + 2s) - 4s\cos(k\Delta x)\xi + 2s - 1 = 0.$$

Therefore,

$$\xi = \frac{2s\cos(k\Delta x) \pm \sqrt{1 - 4s^2\sin(k\Delta x)}}{1 + 2s}.$$

First, consider the case when ξ is real. In this case $\xi \leq 1$ as long as

$$\sqrt{1 - 4s^2\sin^2(k\Delta x)} \leq 1 + 2s(1 - \cos(k\Delta x)).$$

Square both sides. Since

$$(1 + 2s(1 - \cos(k\Delta x)))^2 - (1 - 4s^2\sin^2(k\Delta x)) = (1 - \cos(k\Delta x))(8s^2 + 4s) \geq 0$$

for all s, $\xi \leq 1$ for all s. Similarly it follows that $\xi \geq -1$ for all s.

Next consider the case when ξ is complex. In this case

$$\begin{aligned}
|\xi|^2 &= (\operatorname{Re}(\xi))^2 + (\operatorname{Im}(\xi))^2 \\
&= \frac{(2s\cos(k\Delta x))^2 + 1 - 4s^2\sin^2(k\Delta x)}{(1 + 2s)^2}.
\end{aligned}$$

Therefore, $|\xi|^2 \leq 1$ as long as

$$(2s\cos(k\Delta x))^2 + 1 - 4s^2\sin^2(k\Delta x) \leq (1 + 2s)^2.$$

Since

$$(1 + 2s)^2 - (2s\cos(k\Delta x))^2 - 1 + 4s^2\sin^2(k\Delta x) = 4s(1 + 2s\sin^2(k\Delta x)) \geq 0$$

for all s, we see that this scheme is stable for all s.

8.2.14.

(a) $\dfrac{u_{j,k}^{n+1} - u_{j,k}^n}{\Delta t} = \dfrac{u_{j+1,k}^n - 2u_{j,k}^n + u_{j-1,k}^n}{(\Delta x)^2} + \dfrac{u_{j,k+1}^n - 2u_{j,k}^n + u_{j,k-1}^n}{(\Delta y)^2}$

(b) Separation of variables $u_{j,k}^n = T_n X_j Y_k$ leads to

$$\frac{T_{n+1} - T_n}{T_n} = s_1\frac{X_{j+1} - 2X_j + X_{j-1}}{X_j} + s_2\frac{Y_{k+1} - 2Y_k + Y_{k-1}}{Y_k} = -\alpha - \beta$$

for some constants α and β. By the same reasoning as in the one-dimensional case, it follows that

$$\alpha = 2s_1(1 - \cos(k\Delta x)), \qquad \beta = 2s_2(1 - \cos(l\Delta y)).$$

Letting $\lambda = \alpha + \beta$, we see that $T_{n+1} = (1 - \lambda)T_n$, so stability requires $-1 \le 1 - \lambda \le 1$. Since $\lambda \ge 0$, the second inequality always holds. Thus we only need $\lambda \le 2$ to hold, so we require that

$$2s_1(1 - \cos(k\Delta x)) + 2s_2(1 - \cos(l\Delta y)) \le 2.$$

Since the terms in parentheses can be as large as 2, we need $s_1 + s_2 \le \frac{1}{2}$.

Section 8.3

8.3.2.

(a) Here $s = 4$, so the template is:

$$
\begin{array}{ccc}
 & * & \\
4 & -6 & 4 \\
 & -1 &
\end{array}
$$

Therefore, the solution is:

$$
\begin{array}{llrrrrrrrrr}
n = 3: & 0 & 0 & 16 & -16 & -13 & 38 & -13 & -16 & 16 & 0 & 0 \\
n = 2: & 0 & 0 & 0 & 4 & 2 & -4 & 2 & 4 & 0 & 0 & 0 \\
n = 1: & 0 & 0 & 0 & 0 & 1 & 2 & 1 & 0 & 0 & 0 & 0 \\
n = 0: & 0 & 0 & 0 & 0 & 0 & 0 & 0 & 0 & 0 & 0 & 0
\end{array}
$$

(b) Here $s = 1$, so we have:

$$
\begin{array}{llrrrrrrrrr}
n = 3: & 0 & 0 & 1 & 2 & 2 & 2 & 2 & 2 & 1 & 0 & 0 \\
n = 2: & 0 & 0 & 0 & 1 & 2 & 2 & 2 & 1 & 0 & 0 & 0 \\
n = 1: & 0 & 0 & 0 & 0 & 1 & 2 & 1 & 0 & 0 & 0 & 0 \\
n = 0: & 0 & 0 & 0 & 0 & 0 & 0 & 0 & 0 & 0 & 0 & 0
\end{array}
$$

(c) Clearly (a) is unstable and (b) is stable.

8.3.3

(a) Since $s = 1$, the scheme takes the form $u_j^{n+1} = u_{j+1}^n + u_{j-1}^n - u_j^{n-1}$. The solution is

$$3 \quad \tfrac{66}{25} \quad \tfrac{59}{25} \quad \tfrac{54}{25} \quad \tfrac{51}{25} \quad 2 \quad \tfrac{51}{25} \quad \tfrac{54}{25} \quad \tfrac{59}{25} \quad \tfrac{66}{25} \quad 3$$

$$\tfrac{72}{25} \quad \tfrac{61}{25} \quad \tfrac{52}{25} \quad \tfrac{9}{5} \quad \tfrac{8}{5} \quad \tfrac{37}{25} \quad \tfrac{36}{25} \quad \tfrac{37}{25} \quad \tfrac{8}{5} \quad \tfrac{9}{5} \quad \tfrac{52}{25} \quad \tfrac{61}{25} \quad \tfrac{72}{25}$$

$$\tfrac{73}{25} \quad \tfrac{12}{5} \quad \tfrac{49}{25} \quad \tfrac{8}{5} \quad \tfrac{33}{25} \quad \tfrac{28}{25} \quad 1 \quad \tfrac{24}{25} \quad 1 \quad \tfrac{28}{25} \quad \tfrac{33}{25} \quad \tfrac{8}{5} \quad \tfrac{49}{25} \quad \tfrac{12}{5} \quad \tfrac{73}{25}$$

$$\tfrac{78}{25} \quad \tfrac{63}{25} \quad 2 \quad \tfrac{39}{25} \quad \tfrac{6}{5} \quad \tfrac{23}{25} \quad \tfrac{18}{25} \quad \tfrac{3}{5} \quad \tfrac{14}{25} \quad \tfrac{3}{5} \quad \tfrac{18}{25} \quad \tfrac{213}{25} \quad \tfrac{6}{5} \quad \tfrac{39}{25} \quad 2 \quad \tfrac{63}{25} \quad \tfrac{78}{25}$$

$$\tfrac{87}{25} \quad \tfrac{14}{5} \quad \tfrac{11}{5} \quad \tfrac{42}{25} \quad \tfrac{31}{25} \quad \tfrac{22}{25} \quad \tfrac{3}{5} \quad \tfrac{2}{5} \quad \tfrac{7}{25} \quad \tfrac{6}{25} \quad \tfrac{7}{25} \quad \tfrac{2}{5} \quad \tfrac{3}{5} \quad \tfrac{22}{25} \quad \tfrac{31}{25} \quad \tfrac{42}{25} \quad \tfrac{11}{5} \quad \tfrac{14}{5} \quad \tfrac{87}{25}$$

$$4 \quad \tfrac{81}{25} \quad \tfrac{64}{25} \quad \tfrac{49}{25} \quad \tfrac{36}{25} \quad 1 \quad \tfrac{16}{25} \quad \tfrac{9}{25} \quad \tfrac{4}{25} \quad \tfrac{1}{25} \quad 0 \quad \tfrac{1}{25} \quad \tfrac{4}{25} \quad \tfrac{9}{25} \quad \tfrac{16}{25} \quad 1 \quad \tfrac{36}{25} \quad \tfrac{49}{25} \quad \tfrac{64}{25} \quad \tfrac{81}{25} \quad 4$$

where the first two rows come from applying (8.3.6). The top row corresponds to $t = 1$.

(b) By d'Alembert's formula, the exact solution is

$$u(x,t) = \frac{1}{2}\left((x-t)^2 + (x+t)^2\right) + \frac{1}{2}\int_{x-t}^{x+t} 1\, dy = x^2 + t^2 + t.$$

The numerical solution exactly agrees with this for all points in the domain $|x| \le 2 - t$, $0 \le t \le 1$.

8.3.5. For the Neumann boundary conditions, we use (8.2.14). For the initial conditions, we use (8.3.6). Here $s = 1$ and therefore:

$$n = 7: \quad \left(\tfrac{1}{2}\right) \ 0 \ \tfrac{1}{2} \ 1 \ 1 \ 1 \ \tfrac{1}{2} \ 0 \ \left(\tfrac{1}{2}\right)$$

$$n = 6: \quad (0) \ 0 \ 0 \ 1 \ 2 \ 1 \ 0 \ 0 \ (0)$$

$$n = 5: \quad \left(\tfrac{1}{2}\right) \ 0 \ \tfrac{1}{2} \ 1 \ 1 \ 1 \ \tfrac{1}{2} \ 0 \ \left(\tfrac{1}{2}\right)$$

$$n = 4: \quad (1) \ 1 \ 1 \ \tfrac{1}{2} \ 0 \ \tfrac{1}{2} \ 1 \ 1 \ (1)$$

$$n = 3: \quad (1) \ 2 \ 1 \ 0 \ 0 \ 0 \ 1 \ 2 \ (1)$$

$$n = 2: \quad (1) \ 1 \ 1 \ \tfrac{1}{2} \ 0 \ \tfrac{1}{2} \ 1 \ 1 \ (1)$$

$$n = 1: \quad \left(\tfrac{1}{2}\right) \ 0 \ \tfrac{1}{2} \ 1 \ 1 \ 1 \ \tfrac{1}{2} \ 0 \ \left(\tfrac{1}{2}\right)$$

$$n = 0: \quad (0) \ 0 \ 0 \ 1 \ 2 \ 1 \ 0 \ 0 \ (0)$$

Since the values at $n = 6$ and $n = 7$ equal those at $n = 0$ and $n = 1$, the solution is periodic with a period of 6 time steps. We see the reflection of the maximum value ($= 2$) at the time steps $n = 3, 9, 15, \ldots$

8.3.7. The scheme is

$$u_j^{n+1} = u_{j+1}^n + u_{j-1}^n + (u_j^n)^3 - u_j^{n-1},$$

so the solution is:

$n = 4$:	0.5	1.375	4.877	29.620	95.549	35	95.549	29.620	4.877	1.375	0.5
$n = 3$:	0	$\frac{1}{2}$	$\frac{5}{4}$	$\frac{1497}{512}$	$\frac{9}{2}$	3	$\frac{9}{2}$	$\frac{1497}{512}$	$\frac{5}{4}$	$\frac{1}{2}$	0
$n = 2$:	0	0	$\frac{1}{2}$	$\frac{9}{8}$	$\frac{3}{2}$	1	$\frac{3}{2}$	$\frac{9}{8}$	$\frac{1}{2}$	0	0
$n = 1$:	0	0	0	$\frac{1}{2}$	1	1	1	$\frac{1}{2}$	0	0	0
$n = 0$:	0	0	0	0	1	2	1	0	0	0	0

The nonlinear term makes it wildly unstable.

8.3.8. With $\Delta x = \Delta t = 1$, the scheme is

$$u_j^{n+1} - 2u_j^n + u_j^{n-1} = u_{j+1}^n - 2u_j^n + u_{j-1}^n - \frac{1}{4}\frac{(u_j^{n+1})^4 - (u_j^{n-1})^4}{u_j^{n+1} - u_j^{n-1}}.$$

This can be simplified to

$$u_j^{n+1} + u_j^{n-1} = u_{j+1}^n + u_{j-1}^n - \frac{1}{4}\left((u_j^{n+1})^3 + (u_j^{n+1})^2 u_j^{n-1} + u_j^{n+1}(u_j^{n-1})^2 + (u_j^{n-1})^3\right).$$

The solution is:

0.431	0.670	0.069	-1.139	-0.718	-0.549	-0.718	-1.139	0.069	0.670	0.431
0	0.451	0.745	0.123	-1.183	-0.723	-1.183	0.123	0.745	0.451	0
0	0	0.473	0.848	0.191	-1.140	0.191	0.848	0.473	0	0
0	0	0	$\frac{1}{2}$	1	1	1	$\frac{1}{2}$	0	0	0
0	0	0	0	1	2	1	0	0	0	0

8.3.9. Multiplying the left hand side of equation (8.3.12) by $v_j^{n+1} - v_j^{n-1}$ gives

$$\frac{v_j^{n+1} - 2v_j^n + v_j^{n-1}}{(\Delta t)^2}(v_j^{n+1} - v_j^{n-1}) = \frac{((v_j^{n+1} - v_j^n) - (v_j^n - v_j^{n-1}))((v_j^{n+1} - v_j^n) + (v_j^n - v_j^{n-1}))}{(\Delta t)^2}.$$

Writing the right hand side as a difference of squares and summing over j gives

$$\sum_j \left(\frac{v_j^{n+1} - v_j^n}{\Delta t}\right)^2 - \left(\frac{v_j^n - v_j^{n-1}}{\Delta t}\right)^2 \equiv A.$$

Next, multiplying the first term on the right hand side of (8.3.12) by $v_j^{n+1} - v_j^{n-1}$ gives

$$\frac{v_{j+1}^n - 2v_j^n + v_{j-1}^n}{(\Delta r)^2}(v_j^{n+1} - v_j^{n-1}) = \frac{((v_{j+1}^n - v_j^n) - (v_j^n - v_{j-1}^n))(v_j^{n+1} - v_j^{n-1})}{(\Delta r)^2},$$

so summing over j gives

$$\frac{1}{(\Delta r)^2}\left[\sum_j (v_{j+1}^n - v_j^n)(v_j^{n+1} - v_j^{n-1}) - \sum_j (v_j^n - v_{j-1}^n)(v_j^{n+1} - v_j^{n-1})\right].$$

Since v vanishes at $r = 0$, if we assume that v vanishes as $r \to +\infty$, it follows that we may replace the index j in either sum with $j + 1$. Doing this in the second sum gives

$$\frac{1}{(\Delta r)^2}\left[\sum_j (v_{j+1}^n - v_j^n)(v_j^{n+1} - v_j^{n-1}) - \sum_j (v_{j+1}^n - v_j^n)(v_{j+1}^{n+1} - v_{j+1}^{n-1})\right],$$

which may be rewritten

$$-\frac{1}{(\Delta r)^2}\left[\sum_j (v_{j+1}^{n+1} - v_j^{n+1})(v_{j+1}^n - v_j^n) - \sum_j (v_{j+1}^n - v_j^n)(v_{j+1}^{n-1} - v_j^{n-1})\right] \equiv B.$$

Next, multiplying the second term on the right side of (8.3.12) by $v_j^{n+1} - v_j^{n-1}$ and summing gives

$$-\frac{1}{2}\sum_j (v_j^{n+1} + v_j^{n-1})(v_j^{n+1} - v_j^{n-1}) = -\frac{1}{2}\sum_j [(v_j^{n+1})^2 - (v_j^{n-1})^2] \equiv C.$$

Finally, multiplying the third term on the right side of (8.3.12) by $v_j^{n+1} - v_j^{n-1}$ and summing gives

$$-\frac{1}{8}\sum_j \frac{(v_j^{n+1})^8 - (v_j^{n-1})^8}{(j\Delta r)^6} \equiv D.$$

Thus by (8.3.12) we have $A = B + C + D$, which implies $E_n = E_{n-1}$ and therefore E is independent of n.

Section 8.4

8.4.2.

$$
\begin{matrix}
0 & 0 & 0 & 0 \\
0 & 0 & 0 & 24 \\
0 & 0 & 0 & 0 \\
0 & 0 & 0 & 0
\end{matrix}
\longrightarrow
\begin{matrix}
0 & 0 & 0 & 0 \\
0 & 0 & 6 & 24 \\
0 & 0 & 0 & 0 \\
0 & 0 & 0 & 0
\end{matrix}
\longrightarrow
\begin{matrix}
0 & 0 & 0 & 0 \\
0 & \frac{3}{2} & 6 & 24 \\
0 & 0 & \frac{3}{2} & 0 \\
0 & 0 & 0 & 0
\end{matrix}
$$

$$
\longrightarrow
\begin{matrix}
0 & 0 & 0 & 0 \\
0 & \frac{3}{2} & \frac{27}{4} & 24 \\
0 & \frac{3}{4} & \frac{3}{2} & 0 \\
0 & 0 & 0 & 0
\end{matrix}
\longrightarrow
\begin{matrix}
0 & 0 & 0 & 0 \\
0 & \frac{15}{8} & \frac{27}{4} & 24 \\
0 & \frac{3}{4} & \frac{15}{8} & 0 \\
0 & 0 & 0 & 0
\end{matrix}
$$

$$
\longrightarrow
\begin{matrix}
0 & 0 & 0 & 0 \\
0 & \frac{15}{8} & \frac{111}{16} & 24 \\
0 & \frac{15}{16} & \frac{15}{8} & 0 \\
0 & 0 & 0 & 0
\end{matrix}
\longrightarrow
\begin{matrix}
0 & 0 & 0 & 0 \\
0 & \frac{63}{32} & \frac{111}{16} & 24 \\
0 & \frac{15}{16} & \frac{63}{32} & 0 \\
0 & 0 & 0 & 0
\end{matrix}
$$

8.4.3.

$$
\begin{matrix}
0 & 0 & 0 & 0 \\
0 & 0 & 0 & 24 \\
0 & 0 & 0 & 0 \\
0 & 0 & 0 & 0
\end{matrix}
\longrightarrow
\begin{matrix}
0 & 0 & 0 & 0 \\
0 & 0 & 6 & 24 \\
0 & 0 & 0 & 0 \\
0 & 0 & 0 & 0
\end{matrix}
\longrightarrow
\begin{matrix}
0 & 0 & 0 & 0 \\
0 & \frac{3}{2} & \frac{27}{4} & 24 \\
0 & 0 & \frac{3}{2} & 0 \\
0 & 0 & 0 & 0
\end{matrix}
$$

$$
\longrightarrow
\begin{matrix}
0 & 0 & 0 & 0 \\
0 & \frac{15}{8} & \frac{111}{16} & 24 \\
0 & \frac{3}{4} & \frac{15}{8} & 0 \\
0 & 0 & 0 & 0
\end{matrix}
\longrightarrow
\begin{matrix}
0 & 0 & 0 & 0 \\
0 & \frac{63}{32} & \frac{447}{64} & 24 \\
0 & \frac{15}{16} & \frac{63}{32} & 0 \\
0 & 0 & 0 & 0
\end{matrix}
$$

8.4.5. The boundary values are $u_{1,0} = 24$, $u_{2,0} = 48$ and $u_{j,k} = 0$ elsewhere.

$$
\begin{aligned}
u_{1,1} &= \tfrac{1}{4}(u_{2,1} + u_{1,2} + 24) \\
u_{2,1} &= \tfrac{1}{4}(u_{1,1} + u_{2,2} + 48) \\
u_{1,2} &= \tfrac{1}{4}(u_{1,1} + u_{2,2}) \\
u_{2,2} &= \tfrac{1}{4}(u_{2,1} + u_{1,2})
\end{aligned}
\qquad
\begin{bmatrix} 4 & -1 & -1 & 0 \\ -1 & 4 & 0 & -1 \\ -1 & 0 & 4 & -1 \\ 0 & -1 & -1 & 4 \end{bmatrix}
\begin{bmatrix} u_{1,1} \\ u_{2,1} \\ u_{1,2} \\ u_{2,2} \end{bmatrix}
=
\begin{bmatrix} 24 \\ 48 \\ 0 \\ 0 \end{bmatrix}
$$

Solving this gives $u_{1,1} = 11$, $u_{2,1} = 16$, $u_{1,2} = 4$ and $u_{2,2} = 5$.

$$
\begin{array}{cccc}
0 & 0 & 0 & 0 \\
0 & 4 & 5 & 0 \\
0 & 11 & 16 & 0 \\
0 & 24 & 48 & 0
\end{array}
$$

8.4.6.

(a) $\dfrac{u_{j+1,k} - 2u_{j,k} + u_{j-1,k}}{(\Delta x)^2} + \dfrac{u_{j,k+1} - 2u_{j,k} + u_{j,k-1}}{(\Delta y)^2} = f_{j,k}$

(b)
$$
\begin{array}{ccc}
0 & 0 & 0 \\
0 & -\frac{1}{16} & 0 \\
0 & 0 & 0
\end{array}
$$

(c) Solving the system of equations

$$
\begin{aligned}
u_{1,1} &= \tfrac{1}{4}(u_{2,1} + u_{1,2}) - \tfrac{1}{36} \\
u_{2,1} &= \tfrac{1}{4}(u_{1,1} + u_{2,2}) - \tfrac{1}{36} \\
u_{1,2} &= \tfrac{1}{4}(u_{1,1} + u_{2,2}) - \tfrac{1}{36} \\
u_{2,2} &= \tfrac{1}{4}(u_{2,1} + u_{1,2}) - \tfrac{1}{36} \,,
\end{aligned}
\qquad
\begin{bmatrix} 4 & -1 & -1 & 0 \\ -1 & 4 & 0 & -1 \\ -1 & 0 & 4 & -1 \\ 0 & -1 & -1 & 4 \end{bmatrix}
\begin{bmatrix} u_{1,1} \\ u_{2,1} \\ u_{1,2} \\ u_{2,2} \end{bmatrix}
=
\begin{bmatrix} -\frac{1}{9} \\ -\frac{1}{9} \\ -\frac{1}{9} \\ -\frac{1}{9} \end{bmatrix}
$$

gives $u_{1,1} = u_{2,1} = u_{1,2} = u_{2,2} = -\tfrac{1}{18}$.

$$
\begin{array}{cccc}
0 & 0 & 0 & 0 \\
0 & -\frac{1}{18} & -\frac{1}{18} & 0 \\
0 & -\frac{1}{18} & -\frac{1}{18} & 0 \\
0 & 0 & 0 & 0
\end{array}
$$

(d) First let $v(x,y) = u(x,y) - (x^2 - x)/2$. Then v satisfies $\Delta v = 0$ with boundary data $v(x,0) = v(x,1) = (x - x^2)/2$, $v(0,y) = v(1,y) = 0$. Let w be the solution of $\Delta w = 0$

with boundary data $w(x,0) = w(0,y) = w(1,y) = 0$, $w(x,1) = (x - x^2)/2$. Then by symmetry $v(1/2, 1/2) = 2w(1/2, 1/2)$. The solution w is given as a series by

$$w(x,y) = \sum_{n=1}^{\infty} B_n \sin n\pi x \sinh n\pi y$$

where

$$B_n = \frac{2}{\sinh n\pi} \int_0^1 \frac{x - x^2}{2} \sin n\pi x \, dx = \begin{cases} 0 & n \text{ even} \\ \frac{4}{n^3 \pi^3 \sinh n\pi} & n \text{ odd.} \end{cases}$$

At the center, we have

$$w(1/2, 1/2) = \sum_{n \text{ odd}} \frac{4 \sin(n\pi/2) \sinh(n\pi/2)}{n^3 \pi^3 \sinh n\pi} \approx 0.02566,$$

so

$$u(1/2, 1/2) = v(1/2, 1/2) - (1/2 - (1/2)^2)/2 = 2w(1/2, 1/2) - 1/8 \approx -0.07367.$$

The value at the center from the approximation in part (b) is -0.0625.

8.4.8. For $0 \le j, k \le N$ the scheme is

$$\frac{u_{j+1,k} - 2u_{j,k} + u_{j-1,k}}{(\Delta x)^2} + \frac{u_{j,k+1} - 2u_{j,k} + u_{j,k-1}}{(\Delta y)^2} = f_{j,k}$$

and the ghost points satisfy

$$\frac{u_{N+1,k} - u_{N-1,k}}{2\Delta x} = g_{N,k}$$

$$\frac{u_{-1,k} - u_{1,k}}{2\Delta x} = g_{0,k}$$

for $0 \le k \le N$ and

$$\frac{u_{j,N+1} - u_{j,N-1}}{2\Delta y} = g_{j,N}$$

$$\frac{u_{j,-1} - u_{j,1}}{2\Delta y} = g_{j,0}$$

for $0 \le j \le N$.

8.4.9. In the following, the ghost points are indicated at the four edges.

$$
\begin{array}{cccccc}
\tfrac{2}{3} & \tfrac{2}{3} & \tfrac{2}{3} & \tfrac{2}{3} & & \\[4pt]
0 & 0 & 0 & 0 & 0 & -\tfrac{2}{3} \\[4pt]
0 & 0 & 0 & 0 & 0 & -\tfrac{2}{3} \\[4pt]
0 & 0 & 0 & 0 & 0 & -\tfrac{2}{3} \\[4pt]
0 & 0 & 0 & 0 & 0 & -\tfrac{2}{3} \\[4pt]
 & 0 & 0 & 0 & 0 &
\end{array}
\qquad\longrightarrow\qquad
\begin{array}{cccccc}
\tfrac{2}{3} & \tfrac{2}{3} & \tfrac{2}{3} & \tfrac{43}{96} & & \\[4pt]
\tfrac{5}{24} & \tfrac{1}{6} & \tfrac{5}{24} & \tfrac{7}{32} & 0 & -\tfrac{43}{96} \\[4pt]
0 & 0 & 0 & 0 & -\tfrac{7}{32} & -\tfrac{2}{3} \\[4pt]
0 & 0 & 0 & 0 & -\tfrac{5}{24} & -\tfrac{2}{3} \\[4pt]
0 & 0 & 0 & 0 & -\tfrac{1}{6} & -\tfrac{2}{3} \\[4pt]
 & 0 & 0 & 0 & -\tfrac{5}{24} &
\end{array}
$$

$$
\longrightarrow\qquad
\begin{array}{cccccc}
\tfrac{17}{24} & \tfrac{35}{48} & \tfrac{2}{3} & \tfrac{325}{768} & & \\[4pt]
\tfrac{59}{192} & \tfrac{9}{32} & \tfrac{59}{192} & \tfrac{187}{768} & 0 & -\tfrac{325}{768} \\[4pt]
\tfrac{1}{16} & \tfrac{1}{24} & \tfrac{1}{16} & 0 & -\tfrac{187}{768} & -\tfrac{2}{3} \\[4pt]
0 & 0 & 0 & -\tfrac{1}{16} & -\tfrac{59}{192} & -\tfrac{35}{48} \\[4pt]
0 & 0 & 0 & -\tfrac{1}{24} & -\tfrac{9}{32} & -\tfrac{17}{24} \\[4pt]
 & 0 & 0 & -\tfrac{1}{16} & -\tfrac{59}{192} &
\end{array}
$$

At each step, we have solved for the values in the domain using the Gauss-Seidel procedure, after which we updated the ghost values. Let us compare the values at the point $x = 1$, $y = 2/3$. The approximate $u_{3,2}^{(1)}$ is $-7/32 = -0.219$ and $u_{3,2}^{(2)}$ is $-187/768 = -0.243$. The exact solution at that point is $u(1, 2/3) = -5/18 = -0.278$.

8.4.11. Using the forward difference in t and centered differences in x and y, the scheme for the diffusion equation $v_t = v_{xx} + v_{yy}$ is

$$
\frac{v_{j,k}^{n+1} - v_{j,k}^n}{\Delta t} = \frac{v_{j+1,k}^n - 2v_{j,k}^n + v_{j-1,k}^n}{(\Delta x)^2} + \frac{v_{j,k+1}^n - 2v_{j,k}^n + v_{j,k-1}^n}{(\Delta y)^2}.
$$

Setting $\Delta x = \Delta y$ and $\Delta t = (\Delta x)^2/4$ and solving for $v_{j,k}^{n+1}$ this becomes

$$
v_{j,k}^{n+1} = \frac{1}{4}\left(v_{j+1,k}^n + v_{j-1,k}^n + v_{j,k+1}^n + v_{j,k-1}^n\right),
$$

which is precisely the scheme for Jacobi iteration.

Section 8.5

8.5.1. The PDE takes the form $-\Delta u = f$, where $f = 4$. There is only one interior vertex, so $N = 1$ and therefore the approximate solution is $u_1(x, y) = U_1 v_1(x, y)$ and the system of equations (8.5.5) is the single equation $m_{11} U_1 = f_1$, where

$$m_{11} = \iint_D \nabla v_1 \cdot \nabla v_1 \, dx \, dy \qquad \text{and} \qquad f_1 = \iint_D 4 v_1 \, dx \, dy.$$

Setting $v_1(x, y) = ax + by + c$ and solving $v_1(0,0) = v_1(1,0) = v_1(1,1) = v_1(0,1) = 0$ and $v_1(\frac{1}{2}, \frac{1}{2}) = 1$ within each triangle, it follows that v_1 is given by the formulas shown in Figure 12. Therefore $\nabla v_1 \cdot \nabla v_1 = 4$ on each triangle, so $m_{11} = \iint_D 4 \, dx \, dy = 4$. By the symmetry

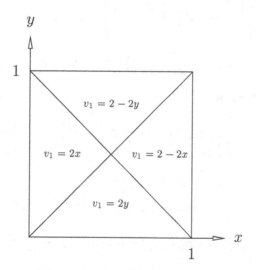

Figure 12: Formulas for v_1.

of v_1 and the square,

$$f_1 = \iint_D 4 v_1 \, dy \, dx = 8 \int_0^{1/2} \int_0^x 4 v_1(x, y) \, dy \, dx = \int_0^{1/2} \int_0^x 64 y \, dy \, dx = \frac{4}{3}.$$

Thus $U_1 = \frac{1}{3}$, so $u_1(1/2, 1/2) = U_1 v_1(1/2, 1/2) = U_1 = \frac{1}{3}$.

8.5.2.

(a) The triangle is half of the parallelogram generated by the vectors $(x_2 - x_1, y_2 - y_1)$ and $(x_3 - x_1, y_3 - y_1)$. Since the area of a parallelogram is the absolute value of the determinant of the matrix whose columns are the generating vectors, the area of the triangle is

$$A = \frac{1}{2}|(x_2 - x_1)(y_3 - y_1) - (x_3 - x_1)(y_2 - y_1)| = \frac{1}{2}|x_2 y_3 - x_1 y_3 - x_2 y_1 - x_3 y_2 + x_1 y_2 + x_3 y_1|.$$

Another method is to take half the length of the cross product of the two vectors.

103

(b) Set $v(x,y) = ax + by + c$ and solve $v(x_1, y_1) = 1$, $v(x_2, y_2) = v(x_3, y_3) = 0$. This leads to the system

$$ax_1 + by_1 + c = 1$$
$$ax_2 + by_2 + c = 0$$
$$ax_3 + by_3 + c = 0.$$

Solving the first equation for c and plugging into the second and third equations gives

$$(x_2 - x_1)a + (y_2 - y_1)b = -1$$
$$(x_3 - x_1)a + (y_3 - y_1)b = -1.$$

Using the formula $\begin{bmatrix} a & b \\ c & d \end{bmatrix}^{-1} = \frac{1}{ad-bc} \begin{bmatrix} d & -b \\ -c & a \end{bmatrix}$, and writing

$$D = (x_2 - x_1)(y_3 - y_1) - (x_3 - x_1)(y_2 - y_1) = x_2 y_3 - x_1 y_3 - x_2 y_1 - x_3 y_2 + x_1 y_2 + x_3 y_1,$$

the solution of this system is

$$\begin{bmatrix} a \\ b \end{bmatrix} = \frac{1}{D} \begin{bmatrix} y_3 - y_1 & y_1 - y_2 \\ x_1 - x_3 & x_2 - x_1 \end{bmatrix} \begin{bmatrix} -1 \\ -1 \end{bmatrix} = \frac{1}{D} \begin{bmatrix} y_2 - y_3 \\ x_3 - x_2 \end{bmatrix},$$

and consequently

$$c = 1 - ax_1 - by_1 = \frac{x_2 y_3 - x_3 y_2}{D}.$$

Thus $v(x,y) = \frac{1}{D}((y_2 - y_3)x + (x_3 - x_2)y + x_2 y_3 - x_3 y_2)$. Another method is to define

$$v(x,y) = \frac{1}{|D|} \left| \begin{bmatrix} x_2 - x \\ y_2 - y \end{bmatrix} \times \begin{bmatrix} x_3 - x \\ y_3 - y \end{bmatrix} \right|.$$

8.5.5. Let $v(x,y) = a + bx + cy + dxy$. Setting $v(0,0) = U_1$ gives $a = U_1$. Setting $v(A,0) = U_2$ gives $U_1 + bA = U_2$, so $b = \frac{1}{A}(U_2 - U_1)$. Setting $v(0,B) = U_3$ gives $U_1 + cB = U_3$, so $c = \frac{1}{B}(U_3 - U_1)$. Finally, setting $v(A,B) = U_4$ gives $U_1 + (U_2 - U_1) + (U_3 - U_1) + dAB = U_4$, so $d = \frac{1}{AB}(U_4 - U_3 - U_2 + U_1)$. Thus

$$v(x,y) = U_1 + (U_2 - U_1)x/A + (U_3 - U_1)y/B + (U_4 - U_3 - U_2 + U_1)xy/(AB).$$

Chapter 9

Section 9.1

9.1.1. We look for a solution of $u_{tt} - c\Delta u = 0$ of the form $u(\mathbf{x}, t) = f(\mathbf{k} \cdot \mathbf{x} - ct)$. By the chain rule, we see that $u_{tt} = c^2 f''(\mathbf{k} \cdot \mathbf{x} - ct)$, $u_{xx} = k_1^2 f''(\mathbf{k} \cdot \mathbf{x} - ct)$, etc. Therefore, $u_{tt} - c^2\Delta u = [c^2 - c^2|\mathbf{k}|]f'' = 0$ implies $f'' = 0$ or $|\mathbf{k}| = 1$. Now $f''(\mathbf{k} \cdot \mathbf{x} - ct) = 0$ implies $f = K_1(\mathbf{k} \cdot \mathbf{x} - ct) + K_2$. Therefore, the possible solutions of the form $f(\mathbf{k} \cdot \mathbf{x} - ct)$ are (i) $f = K_1(\mathbf{k} \cdot \mathbf{x} - ct) + K_2$ where $\mathbf{k}$ is arbitrary or (ii) f is an arbitrary function of $\mathbf{k} \cdot \mathbf{x} - ct$ where $|\mathbf{k}| = 1$.

9.1.4

(a) First, we will show that LM is Lorentz. In order to do so, we need to show that $(LM)^{-1} = \Gamma^{\,t}(LM)\Gamma$. Using properties of inverse matrices and the fact that L and M are Lorentz, we see that

$$(LM)^{-1} = M^{-1}L^{-1} = \Gamma^{\,t}M\Gamma\Gamma^{\,t}L\Gamma.$$

Next, using the fact that Γ is its own inverse and that ${}^tM\,{}^tL = {}^t(LM)$, we see that $\Gamma^{\,t}M\Gamma\Gamma^{\,t}L\Gamma = \Gamma^{\,t}M\,{}^tL\Gamma = \Gamma^{\,t}(LM)\Gamma$. Therefore, we conclude that $(LM)^{-1} = \Gamma^{\,t}(LM)\Gamma$.

Next, we will show that L^{-1} is Lorentz. We need to show that $(L^{-1})^{-1} = \Gamma^{\,t}(L^{-1})\Gamma$. First, we are assuming that L is Lorentz. Therefore, $L^{-1} = \Gamma^{\,t}L\Gamma$. Then, using properties of inverse matrices and the fact that $\Gamma^{-1} = \Gamma$, we see that

$$(L^{-1})^{-1} = (\Gamma^{\,t}L\Gamma)^{-1} = \Gamma^{-1}(^tL)^{-1}\Gamma^{-1} = \Gamma(^tL)^{-1}\Gamma = \Gamma^{\,t}(L^{-1})\Gamma.$$

(b) First, we will assume L is Lorentz to show that $m(L\mathbf{v}) = m(\mathbf{v})$. For $\mathbf{v} = (x, y, z, t)$, we notice that $m(\mathbf{v}) = x^2 + y^2 + z^2 - t^2 = \langle \mathbf{v}, \Gamma\mathbf{v} \rangle$ where $\langle \cdot, \cdot \rangle$ denotes the normal dot product in $\mathbf{R}^4$. Using this fact, we see that $m(L\mathbf{v}) = \langle L\mathbf{v}, \Gamma L\mathbf{v} \rangle = \langle \mathbf{v}, {}^tL\Gamma L\mathbf{v} \rangle$. Since L is Lorentz, $L^{-1} = \Gamma^{\,t}L\Gamma$ which implies $I = \Gamma^{\,t}L\Gamma L$. Using the fact that $\Gamma^{-1} = \Gamma$, we see that $\Gamma = {}^tL\Gamma L$. Therefore, we conclude that $m(L\mathbf{v}) = \langle \mathbf{v}, \Gamma\mathbf{v} \rangle = m(\mathbf{v})$.

It remains to show that if $m(L\mathbf{v}) = m(\mathbf{v})$ for all $\mathbf{v} \in \mathbf{R}^4$, then L is Lorentz. Given $\mathbf{v}$ and $\mathbf{w}$ in $\mathbf{R}^4$, we have $m(L(\mathbf{v} - \mathbf{w})) = m(\mathbf{v} - \mathbf{w})$, so from above we have

$$\langle L(\mathbf{v} - \mathbf{w}), \Gamma L(\mathbf{v} - \mathbf{w}) \rangle = \langle \mathbf{v} - \mathbf{w}, \Gamma(\mathbf{v} - \mathbf{w}) \rangle.$$

Expanding both sides gives

$$\langle L\mathbf{v}, \Gamma L\mathbf{v} \rangle + \langle L\mathbf{w}, \Gamma L\mathbf{w} \rangle - \langle L\mathbf{v}, \Gamma L\mathbf{w} \rangle - \langle L\mathbf{w}, \Gamma L\mathbf{v} \rangle = \langle \mathbf{v}, \Gamma\mathbf{v} \rangle + \langle \mathbf{w}, \Gamma\mathbf{w} \rangle - \langle \mathbf{v}, \Gamma\mathbf{w} \rangle - \langle \mathbf{w}, \Gamma\mathbf{v} \rangle.$$

Using the fact that $\langle L\mathbf{v}, \Gamma L\mathbf{v} \rangle = m(L\mathbf{v}) = m(\mathbf{v}) = \langle \mathbf{v}, \Gamma\mathbf{v} \rangle$ (and likewise for $\mathbf{w}$) and the fact that Γ is symmetric, this reduces to $\langle \Gamma L\mathbf{v}, L\mathbf{w} \rangle = \langle \Gamma\mathbf{v}, \mathbf{w} \rangle$. Equivalently $\langle {}^tL\Gamma L\mathbf{v}, \mathbf{w} \rangle = \langle \Gamma\mathbf{v}, \mathbf{w} \rangle$. Letting $\mathbf{v}$ and $\mathbf{w}$ vary over the standard basis in $\mathbf{R}^4$, it follows that the entries of ${}^tL\Gamma L$ and L are identical. That is ${}^tL\Gamma L = \Gamma$. Right-multiplying by Γ and then left-multiplying by L^{-1}, this becomes $\Gamma^{\,t}L\Gamma = L^{-1}$, so L is Lorentz.

(c) First, using the fact that L is Lorentz, we see that $\Gamma L \Gamma {}^t L = I$. Let L_i denote the ith row of L and $\tilde{L}_i$ denote the ith row of $\tilde{L} = \Gamma L$.

By the chain rule, we see that $U_{x_i x_i} = \sum_{j,k=1}^{4} L_{ji} L_{ki} u_{x_j x_k}$. Therefore,

$$U_{xx} + U_{yy} + U_{zz} - U_{tt} = \sum_{j,k=1}^{4} [L_{j1}L_{k1} + L_{j2}L_{k2} + L_{j3}L_{k3} - L_{j4}L_{k4}] u_{x_j x_k}$$

$$= \sum_{j,k=1}^{4} L_j \cdot \tilde{L}_k u_{x_j x_k} = \sum_{j,k=1}^{4} L_j \cdot (\Gamma L)_k u_{x_j x_k}$$

$$= u_{xx} + u_{yy} + u_{zz} - u_{tt},$$

where in the last step we have used the fact that ${}^t L \Gamma L = \Gamma$.

(d) A Lorentz transformation preserves the Lorentz metric $m(\mathbf{v}) = x^2 + y^2 + z^2 - t^2$. Therefore, a Lorentz transformation moves a 4−vector on the level sets of $x^2 + y^2 + z^2 - t^2$, which are hyperboloids. (Note: This is similar to the way in which a rotation matrix preserves the standard distance metric, thus moving vectors on spheres.)

9.1.7 $E = \iiint (u_t^2 + c^2 |\nabla u|^2) \, d\mathbf{x}$ implies $\dfrac{dE}{dt} = \dfrac{1}{2} \iiint_D [2u_t u_{tt} + 2c^2 \nabla u_t \cdot \nabla u] \, d\mathbf{x}$. By the Divergence Theorem, we note that $\iiint_D \nabla \cdot (u_t \nabla u) \, d\mathbf{x} = \iint_S (u_t \nabla u) \cdot \mathbf{n} \, dS$. Therefore,

$\iiint_D \nabla u_t \cdot \nabla u \, d\mathbf{x} = -\iiint_D u_t \Delta u \, d\mathbf{x} + \iint_S (u_t \nabla u) \cdot \mathbf{n} \, dS$. As a result, we see that $\dfrac{dE}{dt} = \iiint_D u_t [u_{tt} - c^2 \Delta u] \, d\mathbf{x} + c^2 \iint_S u_t \dfrac{\partial u}{\partial n} \, dS$. Using the fact that u is a solution of the wave equation and that $\partial u / \partial n = -b \partial u / \partial t$ for $\mathbf{x} \in S$, we conclude that $\dfrac{dE}{dt} = -bc^2 \iint_S u_t^2 \, dS \le 0$ for $b > 0$. Therefore, the energy is decreasing.

9.1.8

(a) Define the energy as $E = \dfrac{1}{2} \iiint [u_t^2 + c^2 |\nabla u|^2 + m^2 u^2] \, d\mathbf{x}$. Therefore, $E_t = \iiint [u_t u_{tt} + c^2 \nabla u \cdot \nabla u_t + m^2 u u_t] \, d\mathbf{x}$. By the Divergence Theorem, $\iiint_D \nabla \cdot (u_t \nabla u) \, d\mathbf{x} = \iint_S (u_t \nabla u) \cdot \mathbf{n} \, dS$. Using the fact that $\nabla \cdot (u_t \nabla u) = [\nabla u_t \cdot \nabla u + u_t \Delta u]$ and assuming that u vanishes as $|x| \to \infty$, we conclude that $\iiint c^2 \nabla u \cdot \nabla u_t \, d\mathbf{x} = -\iiint c^2 u_t \Delta u \, d\mathbf{x}$. Therefore, $E_t = \iiint u_t [u_{tt} - c^2 \Delta u + m^2 u] \, d\mathbf{x} = 0$. Thus, the energy is constant.

(b) We need to show that the value of the solution $u(\mathbf{x}_0, t_0)$ depends only on the values of $\phi(\mathbf{x})$ and $\psi(\mathbf{x})$ in the ball $B = \{|\mathbf{x} - \mathbf{x}_0| \le ct_0\}$. We will do so as follows. Consider two sets of initial data ϕ_1, ψ_1 and ϕ_2, ψ_2. Suppose $\phi_1(\mathbf{x}) = \phi_2(\mathbf{x})$ and $\psi_1(\mathbf{x}) = \psi_2(\mathbf{x})$ for all $\mathbf{x} \in B$. Let v, w be solutions of the Klein-Gordon equation with initial data ϕ_1, ψ_1 and ϕ_2, ψ_2, respectively. We will show that $v(\mathbf{x}_0, t_0) = w(\mathbf{x}_0, t_0)$ (since the initial data is equal in B), and, therefore, the value of the solution depends only on the value of the initial data in B. Let $u = v - w$. Therefore, $u(\mathbf{x}, 0) = \phi_1(\mathbf{x}) - \phi_2(\mathbf{x}) \equiv 0$ for $\mathbf{x} \in B$, and $u_t(\mathbf{x}, 0) = \psi_1(\mathbf{x}) - \psi_2(\mathbf{x}) \equiv 0 \ \mathbf{x} \in B$. We will show that $u(\mathbf{x}_0, t_0) = 0$.

The proof is a generalization of the one in the text. It proceeds as follows. For u a solution of the Klein-Gordon equation, we see that

$$0 = (u_{tt} - c^2 \Delta u + m^2 u)u_t = \left(\frac{1}{2}u_t^2 + \frac{1}{2}c^2|\nabla u|^2 + \frac{1}{2}m^2 u^2 \right)_t - c^2 \nabla \cdot (u_t \nabla u).$$

Consider a solid cone frustum F with top T, side K and bottom B. We will integrate the above identity over F, showing that if $u \equiv 0$ in B, then $u \equiv 0$ on T. Since T is arbitrary, we will conclude that $u(\mathbf{x}_0, t_0) = 0$. Let $\mathbf{n} = (n_t, n_x, n_y, n_z)$ denote the unit outward normal 4-vector on bdy F and $d\dot{x}$ denote the three-dimensional volume element over bdy F. By the Divergence Theorem, we see that

$$0 = \iiiint_F \left[\left(\frac{1}{2}u_t^2 + \frac{1}{2}c^2|\nabla u|^2 + \frac{1}{2}m^2 u^2 \right)_t - c^2 \nabla \cdot (u_t \nabla u) \right] dt \, d\mathbf{x}$$

$$= \iiint_{\text{bdy} F} \left[n_t \left(\frac{1}{2}u_t^2 + \frac{1}{2}c^2|\nabla u|^2 + \frac{1}{2}m^2 u^2 \right) \right.$$
$$\left. - n_x(c^2 u_t u_x) - n_y(c^2 u_t u_y) - n_z(c^2 u_t u_z) \right] dV$$

The boundary of F has three pieces: T, B and K. On the top T, $\mathbf{n} = (n_t, n_x, n_y, n_z) = (1, 0, 0, 0)$. Therefore, the integral becomes

$$\iiint_T \left(\frac{1}{2}u_t^2 + \frac{1}{2}c^2|\nabla u|^2 + \frac{1}{2}m^2 u^2 \right) d\mathbf{x}.$$

On the bottom B, $=(-1, 0, 0, 0)$. Therefore, this piece is simply

$$\iiint_B (-1) \left(\frac{1}{2}u_t^2 + \frac{1}{2}c^2|\nabla u|^2 + \frac{1}{2}m^2 u^2 \right) d\mathbf{x} = - \iiint_B \left(\frac{1}{2}\psi^2 + \frac{1}{2}c^2|\nabla \phi|^2 + \frac{1}{2}m^2 \phi^2 \right) d\mathbf{x}.$$

On the side K, we use formula (9.1.3) for $\mathbf{n}$. We take the normal with the positive sign to get the outward-pointing normal. Using the fact that $t < t_0$, we see that the integral over K is given by

$$\frac{c}{\sqrt{c^2 + 1}} \iiint_K \left[\frac{1}{2}u_t^2 + \frac{1}{2}c^2|\nabla u|^2 + \frac{1}{2}m^2 u^2 + \frac{x - x_0}{cr}(-c^2 u_t u_x) \right.$$
$$\left. + \frac{y - y_0}{cr}(-c^2 u_t u_y) + \frac{z - z_0}{cr}(-c^2 u_t u_z) \right] dV.$$

The last integrand can be written as

$$I = \frac{1}{2}u_t^2 + \frac{1}{2}c^2|\nabla u|^2 + \frac{1}{2}m^2 u^2 - cu_t u_r$$

where u_r denotes the radial derivative. By completing the square, we see that

$$I = \frac{1}{2}(u_t - cu_r)^2 + \frac{1}{2}c^2|\nabla u - u_r\hat{\mathbf{r}}|^2 + \frac{1}{2}m^2u^2,$$

which is clearly positive. Since $\iiint_T + \iiint_B + \iiint_K = 0$, if $\iiint_K \geq 0$, we must have $\iiint_T + \iiint_B \leq 0$. Therefore,

$$\iiint_T \left(\frac{1}{2}u_t^2 + \frac{1}{2}c^2|\nabla u|^2 + \frac{1}{2}m^2u^2 \right) d\mathbf{x} \leq \iiint_B \left(\frac{1}{2}\psi^2 + \frac{1}{2}c^2|\nabla\phi|^2 + \frac{1}{2}m^2\phi^2 \right) d\mathbf{x}.$$

Since we are assuming that the initial data vanishes in B, we have

$$\iiint_T \left(\frac{1}{2}u_t^2 + \frac{1}{2}c^2|\nabla u|^2 + \frac{1}{2}m^2u^2 \right) d\mathbf{x} \leq 0.$$

Since the terms in the integrand on the left are clearly non-negative, they must all be identically zero. We conclude that $u_t = |\nabla u| = u = 0$ on T. Since we can vary T, we conclude that $u \equiv 0$ in all of the solid cone C that lies above B. In particular, $u(\mathbf{x}_0, t_0) = 0$.

Section 9.2

9.2.1. We use equation (6.1.7) to write the Laplacian in spherical coordinates as

$$\Delta_3 = \frac{\partial^2}{\partial r^2} + \frac{2}{r}\frac{\partial}{\partial r} + \frac{1}{r^2\sin\theta}\frac{\partial}{\partial\theta}\sin\theta\frac{\partial}{\partial\theta} + \frac{1}{r^2\sin^2\theta}\frac{\partial^2}{\partial\phi^2}.$$

The average of $u(\mathbf{x}, t)$ on the sphere $\{|\mathbf{x}| = r\}$ is given by

$$\overline{u}(r, t) = \frac{1}{4\pi r^2} \iint_{|\mathbf{x}|=r} u(\mathbf{x}, t)\, dS.$$

We see that $\overline{u}$ is only a function of r and t. Therefore, $\partial\overline{u}/\partial\theta, \partial\overline{u}/\partial\phi = 0$, which implies

$$\Delta_3\overline{u}(r, t) = \left[\frac{\partial^2}{\partial r^2} + \frac{2}{r}\frac{\partial}{\partial r} \right] \left\{ \frac{1}{4\pi r^2} \iint_{|\mathbf{x}|=r} u(\mathbf{x}, t)\, dS \right\}.$$

Next, we see that

$$\overline{\Delta_3 u}(r, t) = \frac{1}{4\pi r^2} \iint_{|\mathbf{x}|=r} \left[\frac{\partial}{\partial r^2} + \frac{2}{r}\frac{\partial}{\partial r} + \frac{1}{r^2\sin\theta}\frac{\partial}{\partial\theta}\sin\theta\frac{\partial}{\partial\theta} + \frac{1}{r^2\sin^2\theta}\frac{\partial^2}{\partial\phi^2} \right] u(\mathbf{x}, t)\, dS.$$

Now we notice that

$$\frac{1}{4\pi r^2} \iint_{|\mathbf{x}|=r} \frac{1}{r^2\sin\theta}\frac{\partial}{\partial\theta}\left[\sin\theta\frac{\partial}{\partial\theta}u \right] dS = \frac{1}{4\pi} \int_0^{2\pi} \int_0^{\pi} \frac{1}{r^2\sin\theta}\frac{\partial}{\partial\theta}\left[\sin\theta\frac{\partial}{\partial\theta}u \right] r^2\sin\theta\, d\theta\, d\phi$$

$$= \frac{1}{4\pi} \int_0^{2\pi} \int_0^{\pi} \frac{\partial}{\partial\theta}\left[\sin\theta\frac{\partial}{\partial\theta}u \right] d\theta\, d\phi$$

$$= \frac{1}{4\pi} \int_0^{2\pi} \left[\sin\theta\frac{\partial u}{\partial\theta} \right]\Big|_0^{\pi} d\phi = 0.$$

Similarly,

$$\frac{1}{4\pi r^2} \iint_{|\mathbf{x}|=r} \frac{1}{r^2 \sin^2\theta} \frac{\partial^2 u}{\partial\phi^2}\, dS = \frac{1}{4\pi} \int_0^{2\pi} \int_0^{\pi} \frac{1}{r^2 \sin^2\theta} \frac{\partial^2 u}{\partial\phi^2} r^2 \sin\theta\, d\theta\, d\phi$$

$$= \frac{1}{4\pi} \int_0^{\pi} \frac{1}{\sin\theta} \int_0^{2\pi} \frac{\partial^2 u}{\partial\phi^2}\, d\phi\, d\theta$$

$$= \frac{1}{4\pi} \int_0^{\pi} \frac{1}{\sin\theta} \frac{\partial u}{\partial\phi}\Big|_0^{2\pi}\, d\theta = 0.$$

Therefore,

$$\overline{\Delta_3 u}(r,t) = \frac{1}{4\pi r^2} \iint_{|\mathbf{x}|=r} \left[\frac{\partial^2}{\partial r^2} + \frac{2}{r}\frac{\partial}{\partial r}\right] u(\mathbf{x},t)\, dS$$

$$= \int_0^{2\pi} \int_0^{\pi} \left[\frac{\partial^2}{\partial r^2} + \frac{2}{r}\frac{\partial}{\partial r}\right] u \sin\theta\, d\theta\, d\phi$$

$$= \left[\frac{\partial^2}{\partial r^2} + \frac{2}{r}\frac{\partial}{\partial r}\right] \int_0^{2\pi} \int_0^{\pi} u \sin\theta\, d\theta\, d\phi$$

$$= \left[\frac{\partial^2}{\partial r^2} + \frac{2}{r}\frac{\partial}{\partial r}\right] \left\{\frac{1}{4\pi r^2} \iint_{|\mathbf{x}|=r} u(\mathbf{x},t)\, dS\right\}$$

$$= \Delta_3 \overline{u}(r,t).$$

9.2.3. By Kirchhoff's formula (9.2.3),

$$u(\mathbf{x}_0, t_0) = \frac{1}{4\pi c^2 t_0} \iint_S y\, dS,$$

where S is the sphere with center $\mathbf{x}_0$ and radius ct_0. We can rewrite this as

$$u(\mathbf{x}_0, t_0) = \frac{1}{4\pi c^2 t_0} \iint_{|\mathbf{x}-\mathbf{x}_0|=ct_0} y\, dS$$

$$= \frac{t_0}{4\pi c^2 t_0^2} \iint_{|\mathbf{x}-\mathbf{x}_0|=ct_0} y\, dS$$

$$= t_0 \cdot [\text{Average value of } y \text{ on } \{|\mathbf{x}-\mathbf{x}_0| = ct_0\}]$$

$$= t_0 y_0.$$

9.2.5. Suppose ϕ and ψ are identically zero outside a sphere of radius R about the origin. Since a three-dimensional wave travels at exactly speed c, the value of the solution u at a point $\mathbf{x}_0$ and a time t_0 depends only on the initial data on a sphere of radius ct_0 about $\mathbf{x}_0$. Therefore, the solution u will vanish when the intersection of the sphere of radius ct_0 about $\mathbf{x}_0$ and the sphere of radius R about the origin is empty. In particular, the solution $u(\mathbf{x}_0, t_0) \equiv 0$ if $|\mathbf{x}_0| - ct_0 > R$ or $|\mathbf{x}_0| - ct_0 < -R$. At a given time $t_0 > R/c$, the wave must vanish outside the sphere of radius $ct_0 + R$ and inside the sphere of radius $ct_0 - R$.

9.2.6.

(a) If $|\mathbf{x}| > R + \rho$, there is no intersection. Also, if $|\mathbf{x}| + \rho < R$, there is no intersection. If $|\mathbf{x}| + R < \rho$, then S lies entirely within the sphere of center $\mathbf{0}$ and radius ρ. In this case, the surface area of the intersection is just the surface area of S which is $4\pi R^2$. It remains to consider the case $|\rho - R| \le |\mathbf{x}| \le R + \rho$. Using spherical coordinates centered at $\mathbf{x}$ with z-axis in the direction away from the origin, we know the surface area is given by

$$R^2 \int_0^{\phi_0} \int_0^{2\pi} \sin \phi \, d\theta \, d\phi,$$

where ϕ_0 satisfies $\rho^2 = |\mathbf{x}|^2 + R^2 - 2|\mathbf{x}|R \cos \phi_0$, using the law of cosines. See Figure 13. Therefore,

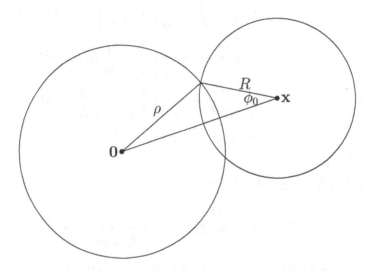

Figure 13: Intersecting spheres when $|\rho - R| \le |\mathbf{x}| \le R + \rho$

$$
\begin{aligned}
S.A. &= 2\pi R^2 \int_0^{\phi_0} \sin \phi \, d\phi \\
&= 2\pi R^2 \left[-\cos \phi |_0^{\phi_0} \right] \\
&= 2\pi R^2 \left[\frac{\rho^2 - |\mathbf{x}|^2 - R^2}{2|\mathbf{x}|R} + 1 \right] \\
&= 2\pi R^2 \left[\frac{\rho^2 - |\mathbf{x}|^2 - R^2 + 2|\mathbf{x}|R}{2|\mathbf{x}|R} \right] \\
&= \frac{\pi R}{|\mathbf{x}|} [\rho^2 - (|\mathbf{x}| - R)^2].
\end{aligned}
$$

(b) By Kirchhoff's formula, we have

$$u(\mathbf{x}_0, t_0) = \frac{1}{4\pi c^2 t_0} \iint_S \psi(\mathbf{x})\, dS$$

$$= \frac{A}{4\pi c^2 t_0} \iint_{S \cap B(0,\rho)} dS.$$

From part (a), if $|\mathbf{x}_0| > ct_0 + \rho$ or $|\mathbf{x}_0| + \rho < ct_0$, the intersection $S \cap B(0,\rho)$ is empty, so that $u(\mathbf{x}_0, t_0) = 0$. Using our result from part (a), for $|\mathbf{x}_0| + ct_0 < \rho$, we have

$$u(\mathbf{x}_0, t_0) = \frac{A}{4\pi c^2 t_0} 4\pi (ct_0)^2 = At_0,$$

while for $|\rho - ct_0| \le |\mathbf{x}_0| \le ct_0 + \rho$,

$$u(\mathbf{x}_0, t_0) = \frac{A}{4\pi c^2 t_0} \frac{\pi ct_0}{|\mathbf{x}_0|} [\rho^2 - (|\mathbf{x}_0| - ct_0)^2]$$

$$= \frac{A}{4c|\mathbf{x}_0|} [\rho^2 - (|\mathbf{x}_0| - ct_0)^2].$$

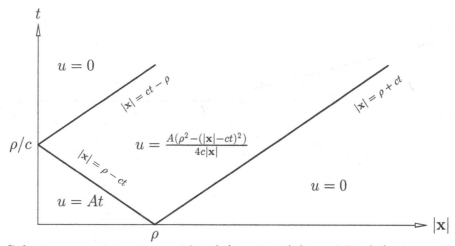

Figure 14: Solution wave equation with $\phi(\mathbf{x}) \equiv 0$, $\psi(\mathbf{x}) = A$ for $|\mathbf{x}| < \rho$ and $\psi(\mathbf{x}) = 0$ for $|\mathbf{x}| > \rho$.

(c) If $\rho = c = A = 1$, then the solution u is given by

$$u(\mathbf{x}, t) = \begin{cases} t & |\mathbf{x}| + t < 1 \\ \frac{[1 - (|\mathbf{x}| - t)^2]}{4|\mathbf{x}|} & |1 - t| \le |\mathbf{x}| \le t + 1. \end{cases}$$

Therefore, after some algebraic simplification, at $t = 1/2$,

$$u(\mathbf{x}, 1/2) = \begin{cases} \frac{1}{2} & |\mathbf{x}| < \frac{1}{2} \\ \frac{3}{16|\mathbf{x}|} - \frac{|\mathbf{x}|}{4} + \frac{1}{4} & \frac{1}{2} \le |\mathbf{x}| \le \frac{3}{2}. \end{cases}$$

At $t = 1$,
$$u(\mathbf{x}, 1) = -\frac{|\mathbf{x}|}{4} + \frac{1}{2} \quad \text{for } 0 \le |\mathbf{x}| \le 2.$$

At $t = 2$,
$$u(\mathbf{x}, 2) = -\frac{|\mathbf{x}|}{4} + 1 - \frac{3}{4|\mathbf{x}|} \quad \text{for } 1 \le |\mathbf{x}| \le 3.$$

These are sketched in Figure 15.

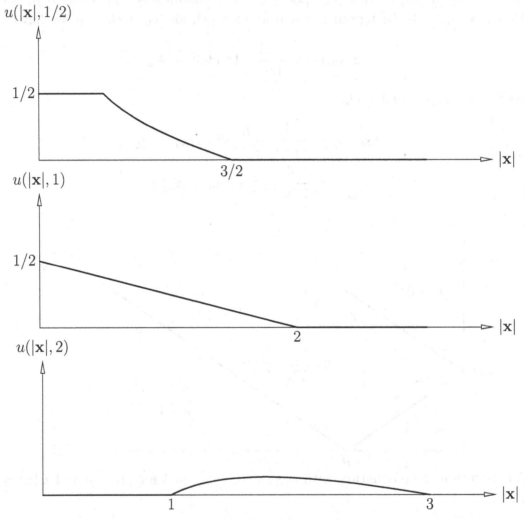

Figure 15: "Movie" of the solution.

(d) For $|\mathbf{x}| = 1/2$, we have

$$u(\mathbf{x}, t) = \begin{cases} t & t < 1/2 \\ \frac{3}{8} + \frac{t}{2} - \frac{t^2}{2} & \frac{1}{2} \le t \le \frac{3}{2}. \end{cases}$$

For $|\mathbf{x}| = 2$, we have

$$u(\mathbf{x}, t) = -\frac{3}{8} + \frac{t}{2} - \frac{t^2}{8} \quad 1 \le t \le 3.$$

These are sketched in Figure 16.

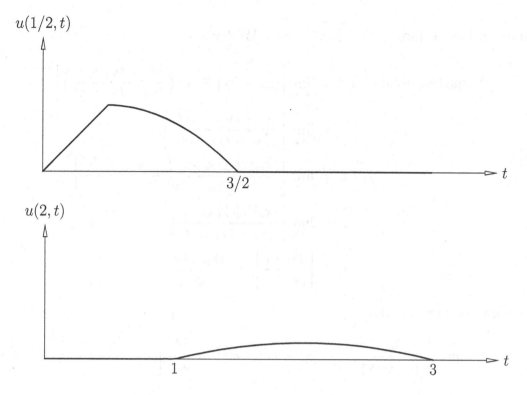

Figure 16: u versus t for $|\mathbf{x}| = \frac{1}{2}$ and $|\mathbf{x}| = 2$.

(e) We need to prove that $t \cdot u(\mathbf{x}_0 + t\mathbf{v}, t)$ converges as $t \to \infty$. From part (b), we know that for $|\mathbf{x}| < \rho$, the solution is given by $u(\mathbf{x}, t) = A[\rho^2 - (|\mathbf{x}| - ct)^2]/4c|\mathbf{x}|$. Therefore, putting $\mathbf{x} = \mathbf{x}_0 + t\mathbf{v}$,

$$t \cdot u(\mathbf{x}_0 + t\mathbf{v}, t) = \frac{At}{4c|\mathbf{x}_0 + t\mathbf{v}|}[\rho^2 - (|\mathbf{x}_0 + t\mathbf{v}| - ct)^2].$$

First, we will look at $\lim_{t \to \infty} At/4c|\mathbf{x}_0 + t\mathbf{v}|$. Using L'Hopital's rule, we see that

$$\lim_{t \to \infty} \frac{At}{4c|\mathbf{x}_0 + t\mathbf{v}|} = \lim_{t \to \infty} \frac{A}{4c}\sqrt{\frac{t^2}{|\mathbf{x}_0|^2 + 2t\mathbf{x}_0 \cdot \mathbf{v} + t^2|\mathbf{v}|^2}}$$

$$= \frac{A}{4c}\sqrt{\frac{1}{|\mathbf{v}|^2}} = \frac{A}{4c^2}.$$

Next, we look at $\lim_{t \to \infty}[\rho^2 - (|\mathbf{x}_0 + t\mathbf{v}| - ct)^2]$. Clearly, $\lim_{t \to \infty} \rho^2 = \rho^2$. Therefore, it

113

remains to look at $\lim_{t\to\infty}(|\mathbf{x}_0 + t\mathbf{v}| - ct)^2$. We have

$$
\begin{aligned}
\lim_{t\to\infty}(|\mathbf{x}_0 + t\mathbf{v}| - ct)^2 &= \lim_{t\to\infty}\left[(|\mathbf{x}_0 + t\mathbf{v}| - ct)\left(\frac{|\mathbf{x}_0 + t\mathbf{v}| + ct}{|\mathbf{x}_0 + t\mathbf{v}| + ct}\right)\right]^2 \\
&= \lim_{t\to\infty}\left[\frac{|\mathbf{x}_0 + t\mathbf{v}|^2 - c^2t^2}{|\mathbf{x}_0 + t\mathbf{v}| + ct}\right]^2 \\
&= \lim_{t\to\infty}\left[\frac{|\mathbf{x}_0|^2 + 2t\mathbf{x}_0\cdot\mathbf{v} + t^2|\mathbf{v}|^2 - c^2t^2}{|\mathbf{x}_0 + t\mathbf{v}| + ct}\right]^2 \\
&= \lim_{t\to\infty}\left[\frac{|\mathbf{x}_0|^2 + 2t\mathbf{x}_0\cdot\mathbf{v}}{|\mathbf{x}_0 + t\mathbf{v}| + ct}\right]^2 \\
&= \left[\frac{2\mathbf{x}_0\cdot\mathbf{v}}{|\mathbf{v}| + c}\right]^2 = \frac{(\mathbf{x}_0\cdot\mathbf{v})^2}{c^2}.
\end{aligned}
$$

Therefore, we conclude that

$$
\lim_{t\to\infty}\frac{At}{4c|\mathbf{x}_0 + t\mathbf{v}|}[\rho^2 - (|\mathbf{x}_0 + t\mathbf{v}| - ct)^2] = \frac{A}{4c^2}\left[\rho^2 - \frac{(\mathbf{x}_0\cdot\mathbf{v})^2}{c^2}\right].
$$

9.2.7.

(a) By Kirchhoff's formula,

$$
u(\mathbf{x}_0, t) = \frac{\partial}{\partial t}\left[\frac{1}{4\pi c^2 t}\iint_S \phi(\mathbf{x})\, dS\right],
$$

where $\phi(\mathbf{x}) = A$ for $|\mathbf{x}| < \rho$, $\phi(\mathbf{x}) = 0$ for $|\mathbf{x}| < \rho$ and S is the sphere with center $\mathbf{x}_0$ and radius ct. The term in brackets above was evaluated in Exercise 9.2.6(b). In particular, for $\phi(\mathbf{x})$ and S as defined above,

$$
\frac{1}{4\pi c^2 t}\iint_S \phi(\mathbf{x})\, dS = \begin{cases} At & r \leq \rho - ct \\ A\left(\frac{\rho^2 - (r-ct)^2}{4cr}\right) & |\rho - ct| \leq r \leq \rho + ct, \end{cases}
$$

and the integral vanishes for $r > \rho + ct$ and $r < ct - \rho$. Differentiating each of these terms with respect to t, we see that our solution is given by

$$
u(\mathbf{x}_0, t) = \begin{cases} A & r \leq \rho - ct \\ \frac{A(r-ct)}{2r} & |\rho - ct| \leq r \leq \rho + ct \end{cases}
$$

and $u(\mathbf{x}_0, t) = 0$ for $r > \rho + ct$ and $r < ct - \rho$.

(b) See Figure b. The solution has discontinuities along the lines $r = \rho - ct$, $r = \rho + ct$, and $r = ct - \rho$.

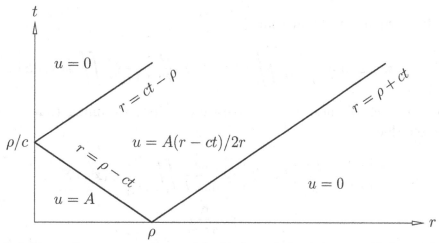

Figure 17: Solution of wave equation with $\phi(\mathbf{x}) = A$ for $|\mathbf{x}| < \rho$, $\phi(\mathbf{x}) = 0$ for $|\mathbf{x}| > \rho$, and $\psi(\mathbf{x}) \equiv 0$.

(c) By part (a), the solution is given by $u(\mathbf{x}, t) = A(|\mathbf{x}| - ct)/2|\mathbf{x}|$. Therefore,

$$
\begin{aligned}
t \cdot u(\mathbf{x}_0 + t\mathbf{v}, t) &= \frac{At(|\mathbf{x}_0 + t\mathbf{v}| - ct)}{2|\mathbf{x}_0 + t\mathbf{v}|} \\
&= \frac{At(|\mathbf{x}_0 + t\mathbf{v}| - ct)(|\mathbf{x}_0 + t\mathbf{v}| + ct)}{2|\mathbf{x}_0 + t\mathbf{v}|(|\mathbf{x}_0 + t\mathbf{v}| + ct)} \\
&= \frac{At(|\mathbf{x}_0 + t\mathbf{v}|^2 - c^2t^2)}{2|\mathbf{x}_0 + t\mathbf{v}|(|\mathbf{x}_0 + t\mathbf{v}| + ct)} \\
&= \frac{At(|\mathbf{x}_0|^2 + 2t\mathbf{x}_0 \cdot \mathbf{v} + t^2|\mathbf{v}|^2 - c^2t^2)}{2(|\mathbf{x}_0 + t\mathbf{v}|^2 + |\mathbf{x}_0 + t\mathbf{v}|ct)} \\
&= \frac{At(|\mathbf{x}_0|^2 + 2t\mathbf{x}_0 \cdot \mathbf{v})}{2(|\mathbf{x}_0 + t\mathbf{v}|^2 + |\mathbf{x}_0 + t\mathbf{v}|ct)} \\
&= \frac{At(|\mathbf{x}_0|^2 + 2t\mathbf{x}_0 \cdot \mathbf{v})}{2(|\mathbf{x}_0|^2 + 2t\mathbf{x}_0 \cdot \mathbf{v} + t^2|\mathbf{v}|^2 + |\mathbf{x}_0 + t\mathbf{v}|ct)}.
\end{aligned}
$$

Therefore,

$$
\lim_{t \to \infty} t \cdot u(\mathbf{x}_0 + t\mathbf{v}, t) = \frac{2A\mathbf{x}_0 \cdot \mathbf{v}}{2|\mathbf{v}|^2 + 2c|\mathbf{v}|} = \frac{A\mathbf{x}_0 \cdot \mathbf{v}}{2c^2}.
$$

9.2.9.

(a) Kirchhoff's formula says that

$$
u(\mathbf{x}_0, t_0) = \frac{1}{4\pi c^2 t_0} \iint_S \psi(\mathbf{x})\, dS + \frac{\partial}{\partial t_0}\left[\frac{1}{4\pi c^2 t_0} \iint_S \phi(\mathbf{x})\, dS \right],
$$

where S is the sphere with center $\mathbf{x}_0$ and radius ct_0. Assume $\phi, \psi \equiv 0$ outside a sphere of radius R about the origin. Let $B(0, R) \equiv \{\mathbf{x} \in \mathbb{R}^3 : |\mathbf{x}| < R\}$. Fix a point $\mathbf{x}_0 = (x_0, y_0, z_0)$. Choose t_0 such that $ct_0 > |\mathbf{x}_0| + R$. Then $B(0, R) \cap S = \emptyset$. Therefore, the above integrals vanish, and $u(\mathbf{x}_0, t_0) = 0$.

(b) By Kirchhoff's formula, we know that

$$t \cdot u(x,y,z,t) = \frac{t}{4\pi c^2 t} \iint_S \psi(\mathbf{x})\, dS + t \frac{\partial}{\partial t}\left[\frac{1}{4\pi c^2 t} \iint_S \phi(\mathbf{x})\, dS \right].$$

We will look at each of these terms separately. First, assuming that $\psi \equiv 0$ outside $B(0,R)$, we see that

$$\left| \frac{t}{4\pi c^2 t} \iint_S \psi(\mathbf{x})\, dS \right| \leq \frac{t}{4\pi c^2 t} \iint_{B(0,R)} |\psi(\mathbf{x})|\, dS$$

$$\leq \frac{K}{4\pi c^2}.$$

Now for the second term for $t \cdot u(x,y,z,t)$, we first need to differentiate the expression with respect to t. Before doing so, we note that the integral depends on t. In order to remove the t from the limits of integration, we make the following change of variables. Let $\widetilde{\mathbf{x}} = (\mathbf{x} - \mathbf{x}_0)/ct$. With this change of variables, we see that

$$\frac{1}{4\pi c^2 t} \iint_S \phi(\mathbf{x})\, dS = \frac{t}{4\pi} \iint_{B(0,1)} \phi(\mathbf{x}_0 + ct\widetilde{\mathbf{x}})\, dS.$$

Then applying the product rule, we see that

$$\frac{\partial}{\partial t}\left[\frac{t}{4\pi} \iint_{B(0,1)} \phi(\mathbf{x}_0 + ct\widetilde{\mathbf{x}})\, dS \right] = \frac{1}{4\pi} \iint_{B(0,1)} \phi(\mathbf{x}_0 + ct\widetilde{\mathbf{x}})\, dS$$

$$+ \frac{t}{4\pi} \iint_{B(0,1)} c\widetilde{\mathbf{x}} \cdot \nabla\phi(\mathbf{x}_0 + ct\widetilde{\mathbf{x}})\, dS.$$

Now reversing the change of variables, we see that

$$\frac{\partial}{\partial t}\left[\frac{t}{4\pi} \iint_{B(0,1)} \phi(\mathbf{x}_0 + ct\widetilde{\mathbf{x}})\, dS \right] = \frac{1}{4\pi c^2 t^2} \iint_S \phi(\mathbf{x})\, dS + \frac{t}{4\pi c^2 t^2} \iint_S \frac{\mathbf{x} - \mathbf{x}_0}{t} \cdot \nabla\phi(\mathbf{x})\, dS.$$

Now multiply each of the terms on right-hand side above by t. Assuming that $\phi \equiv 0$ outside $B(0,R)$, we see that for $t \geq K > 0$,

$$\left| \frac{t}{4\pi c^2 t^2} \iint_S \phi(\mathbf{x})\, dS \right| \leq \frac{1}{4\pi c^2 t} \iint_{B(0,R)} |\phi(\mathbf{x})|\, dS \leq K$$

and

$$\left| \frac{t^2}{4\pi c^2 t^2} \iint_S \frac{\mathbf{x}}{t} \cdot \nabla\phi(\mathbf{x})\, dS \right| \leq \frac{1}{4\pi c^2 t} \iint_{B(0,R)} |\mathbf{x} \cdot \nabla\phi(\mathbf{x})|\, dS \leq K.$$

Therefore, we conclude that $u(x,y,z,t) = O(t^{-1})$ as $t \to \infty$.

9.2.11. If u is a spherical solution of the wave equation, then the solution u will only depend on r and t (not θ or ϕ). Equation (9.2.5) is the wave equation written in spherical coordinates, assuming $u_\phi = u_\theta = 0$. Therefore, in order to find all spherical solutions, we need solve

$$u_{tt} = c^2 u_{rr} + 2c^2 \frac{1}{r} u_r.$$

Multiplying this equation by r and using the product rule for differentiation, we see that

$$(ru)_{tt} = c^2[ru_{rr} + 2u_r] = c^2[(ru_r)_r + u_r] = c^2[(ru)_{rr}].$$

Now let $v = ru$. We see that v satisfies the equation

$$v_{tt} = c^2 v_{rr},$$

which is just the wave equation in one spatial dimension. We know all solutions of this equation are given by $v(r,t) = f(r+ct) + g(r-ct)$ for arbitrary functions f and g. Recalling that $v = ru$, we conclude that all spherical solutions of the wave equation have the form $u(r,t) = [f(r+ct) + g(r-ct)]/r$.

9.2.12. Since we are looking for a solution which depends only on r and t, we look for a solution of the form $u(r,t) = [f(r+ct) + h(r-ct)]/r$. (See Exercise 9.2.11.) In order to satisfy the limiting condition, we need to calculate u_r. By the quotient rule, we see that $u_r = \{r[f'(r+ct) + h'(r-ct)] - [f(r+ct) + h(r-ct)]\}/r^2$. Therefore, $4\pi r^2 u_r = 4\pi[r(f'(r+ct) + h'(r-ct)) - (f(r+ct) + h(r-ct))]$ which implies our limiting condition is given by

$$\lim_{r\to 0} 4\pi r^2 u_r = \lim_{r\to 0} 4\pi[r(f'(r+ct) + h'(r-ct)) - (f(r+ct) + h(r-ct))]$$
$$= \lim_{r\to 0} -4\pi[f(r+ct) + h(r-ct)] = g(t).$$

In addition, we are assuming the solution u has zero initial conditions. Therefore,

$$\lim_{t\to 0}[f(r+ct) + h(r-ct)] = f(r) + h(r) = 0$$
$$\lim_{t\to 0}[cf'(r+ct) - ch'(r-ct)] = cf'(r) - ch'(r) = 0.$$

Combining the above two equalities, we conclude that $f'(r) = 0 = h'(r)$, which implies that $f(r) = K_1$ and $h(r) = K_2$ for $r \geq 0$, but $f(r) + h(r) = 0$ implies $K_1 = -K_2$. Therefore, for $t > 0$, $r - ct > 0$, we see that $u(r,t) = [f(r+ct) + h(r-ct)]/r = [K_1 + K_2]/r = 0$. Now it remains to consider the case $r - ct < 0$. In this case, we make use of the limiting condition. In particular, we see that for $t > 0$,

$$\lim_{r\to 0} -4\pi[f(r+ct) + h(r-ct)] = -4\pi[f(ct) + h(-ct)]$$
$$= K_1 - 4\pi h(-ct) = g(t).$$

First, we note that at $t = 0$, we must have $K_1 - 4\pi h(0) = g(0)$. Using the fact that $h(0) = K_2 = -K_1$ and the assumption that $g(0) = 0$, we see that $K_1 - 4\pi K_2 = K_1 + 4\pi K_1 = 0$. Therefore, we conclude that $K_1 = 0 = K_2$. As a result, we must have $-4\pi h(-ct) = g(t)$ for

$t > 0$. By a change of variables, this requirement may be rewritten as $-4\pi h(t^*) = g(-t^*/c)$ or $h(t^*) = -\dfrac{1}{4\pi} g\left(-\dfrac{t^*}{c}\right)$ for $t^* < 0$. Therefore, for $r - ct < 0$, we have

$$u(r,t) = \frac{f(r+ct) + h(r-ct)}{r} = -\frac{1}{4\pi r} g\left(t - \frac{r}{c}\right).$$

In summary, our solution is given by

$$u(r,t) = \begin{cases} -\frac{1}{4\pi r} g\left(t - \frac{r}{c}\right) & \text{for } r < ct \\ 0 & \text{for } r \geq ct. \end{cases}$$

9.2.13.   We will extend the initial data $\psi(x,y,z)$ to be even about the plane $\{z = 0\}$. Denote this even extension by $\psi_{\text{even}}(\mathbf{x})$. In particular,

$$\psi_{\text{even}}(x,y,z) = \begin{cases} \psi(x,y,z) & z > 0 \\ \psi(x,y,-z) & z < 0. \end{cases}$$

Then, by Kirchhoff's formula,

$$u(\mathbf{x}_0, t_0) = \frac{1}{4\pi c^2 t_0} \iint_S \psi_{\text{even}}(\mathbf{x})\, dS,$$

where S is the sphere with center $\mathbf{x}_0$ and radius ct_0. If $z_0 - ct_0 \geq 0$, then $S \cap \{z < 0\} = \emptyset$, in which case $\psi_{\text{even}}(\mathbf{x}) = \psi(\mathbf{x})$. If $z_0 < ct_0$, then $S \cap \{z < 0\} \neq \emptyset$. In this case,

$$
\begin{aligned}
u(\mathbf{x}_0, t_0) &= \frac{1}{4\pi c^2 t_0} \iint_S \psi_{\text{even}}(\mathbf{x})\, dS \\
&= \frac{1}{4\pi c^2 t_0} \iint_{\{z>0\} \cap S} \psi(\mathbf{x})\, dS + \frac{1}{4\pi c^2 t_0} \iint_{\{z<0\} \cap S} \psi(x,y,-z)\, dS.
\end{aligned}
$$

9.2.16.

(a) Using equation (9.2.19), the solution is given by

$$u(x_0, y_0, t_0) = \iint_{D \cap \{|\mathbf{x}| < \rho\}} \frac{A}{[c^2 t_0^2 - (x - x_0)^2 - (y - y_0)^2]^{1/2}} \frac{dx\, dy}{2\pi c},$$

where D is the disk $\{(x - x_0)^2 + (y - y_0)^2 \leq c^2 t_0^2\}$.

If $|\mathbf{x}_0| - ct_0 > \rho$, then $D \cap \{|\mathbf{x}| < \rho\} = \emptyset$, in which case $u(x_0, y_0, t_0) = 0$. If $ct_0 > |\mathbf{x}_0| + \rho$, then $\{|\mathbf{x}| < \rho\} \subseteq D$. If $|\mathbf{x}_0| + ct_0 < \rho$, then $D \subseteq \{|\mathbf{x}| < \rho\}$.

Otherwise, $D \cap \{|\mathbf{x}| < \rho\}$ is a lens-shaped region.

(b) We now calculate the solution at the origin. That is, we find $u(\mathbf{0}, t)$. We have two cases to consider, $ct_0 \leq \rho$ and $ct_0 > \rho$. In the case $ct_0 \leq \rho$,

$$u(\mathbf{0}, t) = \iint_D \frac{A}{[c^2 t_0^2 - x^2 - y^2]^{1/2}} \frac{dx\, dy}{2\pi c},$$

where D denotes the disk of radius ct_0 about $\mathbf{0}$. To calculate the solution, we use polar coordinates. In particular, we can rewrite

$$u(\mathbf{0}, t_0) = \frac{1}{2\pi c} \int_0^{2\pi} \int_0^{ct_0} \frac{A}{[c^2 t_0^2 - r^2]^{1/2}} r \, dr \, d\theta.$$

Using integration by substitution, we see that

$$u(\mathbf{0}, t_0) = \frac{1}{4\pi c} \int_0^{2\pi} \int_0^{c^2 t_0^2} \frac{A}{u^{1/2}} \, du \, d\theta$$

$$= \frac{1}{4\pi c} \int_0^{2\pi} 2A u^{1/2} \Big|_0^{c^2 t_0^2} \, d\theta = \frac{2A}{4\pi c} \int_0^{2\pi} ct_0 \, d\theta = At_0.$$

It remains to consider the case when $ct_0 > \rho$. Since $\psi \equiv 0$ outside $\{|\mathbf{x}| < \rho\}$, it suffices to integrate ψ over $\{|\mathbf{x}| < \rho\}$. Therefore, in this case, our solution is given by

$$u(\mathbf{0}, t_0) = \iint_{\{|\mathbf{x}| < \rho\}} \frac{A}{[c^2 t_0^2 - x^2 - y^2]^{1/2}} \frac{dx \, dy}{2\pi c}$$

$$= \frac{1}{2\pi c} \int_0^{2\pi} \int_0^{\rho} \frac{A}{[c^2 t_0^2 - r^2]^{1/2}} r \, dr \, d\theta$$

$$= -\frac{1}{4\pi c} \int_0^{2\pi} \int_{c^2 t_0^2}^{c^2 t_0^2 - \rho^2} \frac{A}{u^{1/2}} \, du \, d\theta$$

$$= \frac{1}{2\pi c} \int_0^{2\pi} A[(c^2 t_0^2)^{1/2} - (c^2 t_0^2 - \rho^2)^{1/2}] \, d\theta$$

$$= A[t_0 - (t_0^2 - \rho^2/c^2)^{1/2}].$$

9.2.17. Using Exercise 9.2.16, we calculate

$$\lim_{t \to \infty} t \cdot u(\mathbf{0}, t) = \lim_{t \to \infty} tA \left[t - \left(t^2 - \frac{\rho^2}{c^2} \right)^{1/2} \right]$$

$$= \lim_{t \to \infty} tA \left[t - \left(t^2 - \frac{\rho^2}{c^2} \right)^{1/2} \right] \cdot \left[\frac{t + \left(t^2 - \frac{\rho^2}{c^2} \right)^{1/2}}{t + \left(t^2 - \frac{\rho^2}{c^2} \right)^{1/2}} \right]$$

$$= \lim_{t \to \infty} At \left[\frac{t^2 - \left(t^2 - \frac{\rho^2}{c^2} \right)}{t + \left(t^2 - \frac{\rho^2}{c^2} \right)^{1/2}} \right]$$

$$= \lim_{t \to \infty} \frac{At \frac{\rho^2}{c^2}}{t + \left(t^2 - \frac{\rho^2}{c^2} \right)^{1/2}}.$$

Dividing the numerator and denominator by t, we see this limit is given by $A\rho^2/2c^2$.

9.2.19. The goal is to show that $|u(x_0, y_0, t_0)| \leq Ct_0^{-1/2}$ for all t sufficiently large, for some constant that does not depend on x_0, y_0 or t. We prove this for the first term in equation (9.2.19); – the proof for the second term is similar. First let E be the disk outside of which ψ vanishes, let R be its radius, and let M be the maximum of $|\psi(x, y)|$ on this disk. Then

$$\left| \iint_D \frac{\psi(x, y)}{[c^2t_0^2 - (x - x_0)^2 - (y - y_0)^2]^{1/2}} \frac{dx\, dy}{2\pi c} \right| \leq \iint_{D \cap E} \frac{M}{[c^2t_0^2 - (x - x_0)^2 - (y - y_0)^2]^{1/2}} \frac{dx\, dy}{2\pi c},$$

where D is the disk $(x - x_0)^2 + (y - y_0)^2 \leq c^2t_0^2$. The difficult part of estimating this last integral is that the disk D, and hence $D \cap E$, varies with (x_0, y_0). Since we are interested in large t, we will suppose without loss of generality that $t_0 > \max\{1, 8R/c\}$. We will separate the problem into two cases.

Case 1. $E \subset \{(x, y) \mid [(x - x_0)^2 + (y - y_0)^2]^{1/2} \leq \frac{3}{4}ct_0\}$. In this case,

$$c^2t_0^2 - (x - x_0)^2 - (y - y_0)^2 \geq \frac{7}{16}c^2t_0^2,$$

so

$$\iint_{D \cap E} \frac{M}{[c^2t_0^2 - (x - x_0)^2 - (y - y_0)^2]^{1/2}} \frac{dx\, dy}{2\pi c} \leq \frac{2MR^2}{\sqrt{7}c^2t_0} \leq \frac{2MR^2}{\sqrt{7}c^2t_0^{1/2}},$$

where the last inequality follows from the assumption that $t_0 \geq 1$.

Case 2. $E \not\subset \{(x, y) \mid [(x - x_0)^2 + (y - y_0)^2]^{1/2} \leq \frac{3}{4}ct_0\}$. In this case, since the diameter of E is $2R < \frac{1}{4}ct_0$, it follows that E lies completely *outside* the disk $\{(x, y) \mid [(x - x_0)^2 + (y - y_0)^2]^{1/2} \leq \frac{1}{2}ct_0\}$. See Figure 18. To estimate the integral, we will use polar

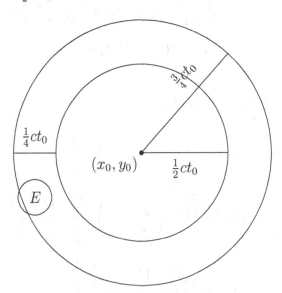

Figure 18: $E \not\subset \{(x, y) \mid [(x - x_0)^2 + (y - y_0)^2]^{1/2} \leq \frac{3}{4}ct_0\}$.

coordinates centered at (x_0, y_0) with $\theta = 0$ defined by the line segment from the center of D to the center of E. The disk E is contained within the polar rectangle $-\theta_0 \leq \theta \leq \theta_0$, $r_0 < r < r_0 + 2R$, as shown in Figure 19. So since the integrand is positive,

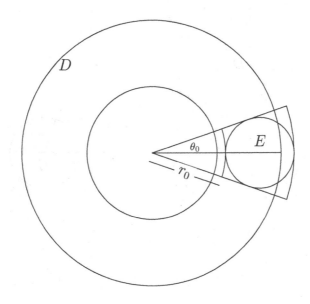

Figure 19: The polar rectangle containing E.

$$\iint_{D \cap E} \frac{M}{[c^2 t_0^2 - (x - x_0)^2 - (y - y_0)^2]^{1/2}} \frac{dx\,dy}{2\pi c} \le \int_{-\theta_0}^{\theta_0} \int_{r_0}^{r_1} \frac{M}{2\pi c[c^2 t_0^2 - r^2]^{1/2}} r\,dr\,d\theta,$$

where $r_1 = \min\{r_0 + 2R, ct_0\}$. Since the integrand is strictly increasing in r, the latter integral may be bounded above by

$$\frac{\theta_0 M}{\pi c} \int_{ct_0 - 2R}^{ct_0} \frac{r}{[c^2 t_0^2 - r^2]^{1/2}}\,dr = \frac{\theta_0 M}{\pi c} \left(2R(2ct_0 - 2R)\right)^{1/2} \le \frac{2\theta_0 M R^{1/2} t_0^{1/2}}{\pi c^{1/2}}.$$

As illustrated in Figure 20, $\tan \theta_0 = R/\ell < 2R/ct_0$ and thus $\theta_0 < \arctan(2R/ct_0) < 2R/ct_0$, so the last expression is bounded by

$$\frac{4M R^{3/2}}{\pi c^{3/2} t_0^{1/2}},$$

as desired.

Section 9.3

9.3.2. Suppose $|\nabla \gamma(\mathbf{x})| = \dfrac{1}{c}$. We need to show that any level surface of $t - \gamma(\mathbf{x})$ is characteristic. We proceed as follows. We consider any curve $\mathbf{x}(t)$ such that $d\mathbf{x}/dt = c^2 \nabla \gamma(\mathbf{x})$. We will show that any such curve $\mathbf{x}(t)$ is a light ray. Then we will show that any level surface of $t - \gamma(\mathbf{x})$ can be written as a union of such light rays. Finally, we will conclude that each of these light rays is orthogonal to the time slices S_t.

Let $\mathbf{x}(t)$ be any curve such that $d\mathbf{x}/dt = c^2 \nabla \gamma(\mathbf{x})$. We will show that $\mathbf{x}(t)$ is a light ray. In order to do so, we need to show that $\mathbf{x}(t)$ is a linear function of t and that $|d\mathbf{x}/dt| = c$. First, we see that $|d\mathbf{x}/dt| = c^2 |\nabla \gamma(\mathbf{x})| = c^2/c = c$. In order to show that $\mathbf{x}(t)$ is a linear

121

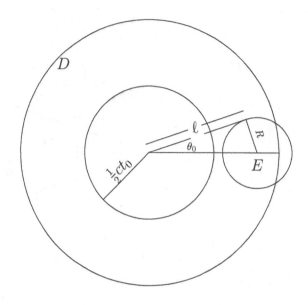

Figure 20: $\tan\theta_0 = R/\ell$

function of t, we will show that $d^2\mathbf{x}/dt^2 = 0$. Let $\gamma_i = \gamma_{x_i}$. By a straightforward calculation, we see that

$$\begin{aligned}
\frac{d^2\mathbf{x}}{dt^2} &= c^2\left(\gamma_{11}\frac{dx_1}{dt} + \gamma_{12}\frac{dx_2}{dt} + \gamma_{13}\frac{dx_3}{dt}, \dots\right) \\
&= c^2\left(\nabla\gamma_1 \cdot \frac{d\mathbf{x}}{dt}, \nabla\gamma_2 \cdot \frac{d\mathbf{x}}{dt}, \nabla\gamma_3 \cdot \frac{d\mathbf{x}}{dt}\right) \\
&= c^4\left(\nabla\gamma_1 \cdot \nabla\gamma, \nabla\gamma_2 \cdot \nabla\gamma, \nabla\gamma_3 \cdot \nabla\gamma\right).
\end{aligned}$$

We are assuming that $|\nabla\gamma(\mathbf{x})| = 1/c$. That is, $\gamma_1^2 + \gamma_2^2 + \gamma_3^2 = (1/c)^2$. Applying the differential operator $\partial/\partial x_i$ to this equation, we see that $2\gamma_1\gamma_{1i} + 2\gamma_2\gamma_{2i} + 2\gamma_3\gamma_{3i} = 0$. Therefore, $\sum_{j=1}^{3}\gamma_{ij}\gamma_j = 0$ for $i = 1, 2, 3$. We can rewrite this equation as $\nabla\gamma \cdot \nabla\gamma_i = 0$ for $i = 1, 2, 3$. That is, $\nabla\gamma$ is perpendicular to $\nabla\gamma_i$ for $i = 1, 2, 3$. Therefore, for $\mathbf{x}(t)$ defined as above, $d^2\mathbf{x}/dt^2 = 0$. That is, $\mathbf{x}(t)$ is a linear function of t. We conclude that such a curve $\mathbf{x}(t)$ is a light ray.

Now we will show that any level surface of $t - \gamma(\mathbf{x})$ can be written as a union of curves of the form $\mathbf{x}(t)$ where $d\mathbf{x}/dt = c^2\nabla\gamma(\mathbf{x})$. Let $(\mathbf{x}_0, t_0)$ be any point on a level surface of $t - \gamma(\mathbf{x})$. Let $\mathbf{x}(t)$ be a curve passing through $(\mathbf{x}_0, t_0)$ such that $d\mathbf{x}/dt = c^2\nabla\gamma(\mathbf{x})$. We will show that $\dfrac{d}{dt}(t - \gamma(\mathbf{x})) = 0$ along such a curve $\mathbf{x}(t)$ and therefore, such a curve lies entirely on this level surface. By a straightforward calculation, we see that

$$\frac{d}{dt}(t - \gamma(\mathbf{x})) = 1 - \nabla\gamma \cdot \frac{d\mathbf{x}}{dt} = 1 - \nabla\gamma \cdot c^2\nabla\gamma = 1 - c^2|\nabla\gamma|^2 = 1 - c^2\left(\frac{1}{c}\right)^2 = 0.$$

Therefore, a level surface of $t - \gamma(\mathbf{x})$ can be written as a union of light rays.

122

It remains only to show that these light rays are orthogonal to the time slices $S_t = \{\mathbf{x} : t - \gamma(\mathbf{x}) = k\}$. But clearly $\nabla\gamma(\mathbf{x})$ is orthogonal to each of these time slices and $d\mathbf{x}/dt = c^2\nabla\gamma(\mathbf{x})$. Therefore, each of the light rays $\mathbf{x}(t)$ is orthogonal to any time slice S_t. We conclude that any level surface of $t - \gamma(\mathbf{x})$ is characteristic.

9.3.3. Use the Green's Theorem method from Section 3.4. As in (3.4.10), we have

$$0 = \iint_D (u_{tt} - c^2 u_{xx})\, dx\, dt,$$

where D is the "triangle" with a curved side and L_1, L_2 are characteristics. See Figure 21. Therefore

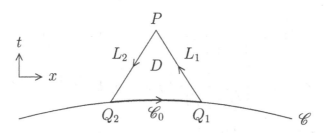

Figure 21: The "triangle" defined by $\mathcal{C}_0$, L_1 and L_2.

$$0 = \int_{\mathcal{C}_0 + L_1 + L_2} (-c^2 u_x\, dt - u_t\, dx).$$

As in Section 3.4,

$$\int_{L_1} = \int_{L_1} c\, du = cu(P) - cu(Q_1)$$

and

$$\int_{L_2} = \int_{L_2} (-c)\, du = cu(P) - cu(Q_2).$$

Let

$$\alpha = \int_{\mathcal{C}_0} (+c^2 u_x\, dt + u_t\, dx).$$

Adding, we get $0 = -\alpha + 2cu(P) - c(u(Q_1) + u(Q_2))$, or

$$u(P) = \frac{1}{2}\left[u(Q_1) + u(Q_2)\right] + \frac{\alpha}{2c}$$

$$= \frac{1}{2}\left[\phi(Q_1) + \phi(Q_2)\right] + \frac{\alpha}{2c}. \tag{S-5}$$

Now $\alpha = \int_{\mathscr{C}_0}$ (linear combination of first derivatives), so α can be expressed uniquely in terms of ϕ and ψ. This uses $c\gamma' \neq 1$. The first initial condition is $u(x, \gamma(x)) = \phi(x)$. Differentiating,

$$u_x + \gamma'(x)u_t = \phi_x. \tag{S-6}$$

The second initial condition is $\frac{\partial u}{\partial n}(x, \gamma(x)) = \psi(x)$. Take $\mathbf{n}$ to be the unit normal vector pointing in the direction of positive time. Then

$$\mathbf{n} = \frac{1}{\sqrt{1 + (\gamma')^2}} \begin{bmatrix} -\gamma' \\ 1 \end{bmatrix}. \tag{S-7}$$

So $\psi = \frac{\partial u}{\partial n} = \frac{1}{\sqrt{1+(\gamma')^2}}[-\gamma'u_x + u_t]$, or $-\gamma'u_x + u_t = \sqrt{1 + (\gamma')^2}\psi$. Solving equations (S-6) and (S-7) for u_x and u_t, we get

$$u_x = \frac{1}{\sqrt{1 + (\gamma')^2}}\left[\phi_x - \gamma'\sqrt{1 + (\gamma')^2}\psi\right]$$

$$u_t = \frac{1}{\sqrt{1 + (\gamma')^2}}\left[\gamma'\phi_x + \sqrt{1 + (\gamma')^2}\psi\right].$$

Along $\mathscr{C}_0$ we have $dt = \gamma'(x)\,dx$. Thus

$$\begin{aligned}
\alpha &= \int_{x_2}^{x_1} \left\{ \frac{c^2\gamma'}{1 + (\gamma')^2}[\phi_x - \gamma'\sqrt{1 + (\gamma')^2}\psi] + \frac{1}{1 + (\gamma')^2}[\gamma'\phi_x + \sqrt{1 + (\gamma')^2}\psi] \right\} dx \\
&= \int_{x_2}^{x_1} \left\{ (1 + c^2)\gamma'\phi_x + (1 - c^2(\gamma')^2)\sqrt{1 + (\gamma')^2}\psi \right\} \frac{dx}{1 + (\gamma')^2},
\end{aligned} \tag{S-8}$$

where we have denoted $Q_1 = (x_1, \gamma(x_1)) = (x_1, t_1)$ and $Q_2 = (x_2, \gamma(x_2)) = (x_2, t_2)$. Now Q_2 is connected to $P = (x_0, t_0)$ be a line of slope $\frac{dx}{dt} = c$. Thus $x_0 - x_2 = c[t_0 - \gamma(x_2)]$. Similarly Q_1 is connected to P by a line of slope $\frac{dx}{dt} = -c$, so that $x_0 - x_1 = -c[t_0 - \gamma(x_1)]$. These two equations determine x_2 and x_1 implicitly. They can be solved because $c|\gamma'(x)| \neq 1$, although they cannot in general be solved explicitly. Putting (S-8) into (S-5) we get the formula

$$u(P) = \frac{1}{2}[\phi(Q_1) + \phi(Q_2)] + \frac{1}{2c}\int_{x_2}^{x_1} \left\{ (1 + c^2)\gamma'\phi_x + (1 - c^2(\gamma')^2)\sqrt{1 + (\gamma')^2}\psi \right\} \frac{dx}{1 + (\gamma')^2}.$$

Note that this formula reduces to equation (2.1.8) if $\mathscr{C}$ is the line $\{t = 0\}$.

9.3.6. Let $u(\mathbf{x}, t) = \int_0^t \mathscr{S}(t - s)f(\mathbf{x}, s)\,ds$, where

$$\mathscr{S}(t)\psi(\mathbf{x}_0) = \frac{1}{4\pi c^2 t} \iint_S \psi(\mathbf{x})\,dS_\xi$$

and $S = \{|\xi - \mathbf{x}_0| = ct\}$. We need to show that u satisfies the inhomogeneous wave equation with zero initial data. First, we will show that u defined above satisfies zero initial conditions.

Clearly, $u(\mathbf{x}, 0) = 0$. Second,

$$u_t(\mathbf{x}, t) = \mathscr{S}(0)f(\mathbf{x}, t) + \int_0^t \frac{\partial}{\partial t}[\mathscr{S}(t-s)f(\mathbf{x}, s)]\, ds.$$

Now for any given function $\psi(\mathbf{x})$, $\mathscr{S}(0)\psi(\mathbf{x}) \equiv \lim_{t \to 0^+} \mathscr{S}(t)\psi(\mathbf{x})$. But,

$$\mathscr{S}(t)\psi(\mathbf{x}_0) = \frac{t}{4\pi c^2 t^2} \iint_S \psi(\xi)\, dS_\xi$$
$$= t \cdot (\text{avg. of } \psi \text{ on } \{|\xi - \mathbf{x}_0| = ct\}).$$

As $t \to 0$, (the average of ψ on $\{|\xi - \mathbf{x}_0| = ct\}) \to \psi(\mathbf{x}_0)$. Therefore, as $t \to 0$, $\mathscr{S}(t)\psi(\mathbf{x}_0) \to 0$. Substituting $\psi(\mathbf{x}) = f(\mathbf{x}, 0)$, we see that

$$u_t(\mathbf{x}, 0) = \mathscr{S}(0)f(\mathbf{x}, 0) + \int_0^0 \cdots ds = 0.$$

Now we will show that u defined above satisfies the wave equation. From above, we calculated that

$$u_t(\mathbf{x}, t) = \int_0^t \frac{\partial}{\partial t}[\mathscr{S}(t-s)f(\mathbf{x}, s)]\, ds.$$

Therefore, by the chain rule,

$$u_{tt} = \frac{\partial}{\partial t}[\mathscr{S}(t-s)f(\mathbf{x}, s)]\Big|_{s=t} + \int_0^t \frac{\partial^2}{\partial t^2}[\mathscr{S}(t-s)f(\mathbf{x}, s)]\, ds.$$

Also,

$$c^2 \Delta u = \int_0^t c^2 \Delta[\mathscr{S}(t-s)f(\mathbf{x}, s)]\, ds.$$

Since $\mathscr{S}(t)\psi(\mathbf{x})$ is a solution of the homogeneous wave equation for any function ψ, we see that

$$u_{tt} - c^2 \Delta u = \frac{\partial}{\partial t}[\mathscr{S}(t-s)f(\mathbf{x}, s)]\Big|_{s=t} + \int_0^t \left(\frac{\partial^2}{\partial t^2} - c^2 \Delta \right)[\mathscr{S}(t-s)f(\mathbf{x}, s)]\, ds$$
$$= \frac{\partial}{\partial t}[\mathscr{S}(t-s)f(\mathbf{x}, s)]\Big|_{s=t}.$$

Let $S = \{|\xi - \mathbf{x}_0| = c(t-s)\}$. Let $S' = \{|\xi| = 1\}$. By the change of variables, $\xi' = [\xi - \mathbf{x}_0]/[c(t-s)]$, we have

$$\mathscr{S}(t-s)f(\mathbf{x}, s) = \frac{1}{4\pi c^2(t-s)} \iint_{\text{bdy } S} f(\xi, s)\, dS_\xi$$
$$= \frac{(t-s)}{4\pi} \iint_{\text{bdy } S'} f(\mathbf{x}_0 + c(t-s)\xi', s)\, dS_{\xi'}.$$

Therefore,

$$\partial_t(\mathscr{S}(t-s)f(\mathbf{x},s)) = \frac{1}{4\pi} \iint_{\text{bdy } S'} f(\mathbf{x}_0 + c(t-s)\xi', s)\, dS_{\xi'}$$

$$+ \frac{(t-s)}{4\pi} \iint_{\text{bdy } S'} c\xi' \cdot \nabla_\xi f(\mathbf{x}_0 + c(t-s)\xi', s)\, dS_{\xi'}$$

$$= \frac{1}{4\pi c^2(t-s)^2} \iint_{\text{bdy } S} f(\xi, s)\, dS_\xi$$

$$+ \frac{(t-s)}{4\pi c^2(t-s)^2} \iint_{\text{bdy } S} \frac{\xi - \mathbf{x}_0}{t-s} \cdot \nabla_\xi f(\xi, s)\, dS_\xi$$

$$= \frac{1}{4\pi c^2(t-s)^2} \iint_{\text{bdy } S} f(\xi, s)\, dS_\xi$$

$$+ \frac{(t-s)}{4\pi c^2(t-s)^2} \iint_{\text{bdy } S} \frac{\partial f}{\partial n}\, dS_\xi$$

$$= (\text{avg. of } f \text{ on bdy } S) + (t-s) \cdot (\text{avg. of } \partial f/\partial n \text{ on bdy } S).$$

As $s \to t^-$, the sphere bdy S shrinks to the point $\mathbf{x}_0$. Therefore, as $s \to t^-$, $\partial_t(\mathscr{S}(t-s)f(\mathbf{x},s)) \to f(\mathbf{x}_0, t)$. Therefore, $u_{tt} - c^2\Delta u = f$ as claimed.

9.3.7. Solution 1. Using equation (9.3.11) we have

$$u(\mathbf{x}, t) = \int_0^t \mathscr{S}(t-s)f(\mathbf{x})\, ds = \int_0^t \mathscr{S}(\sigma)f(\mathbf{x})\, d\sigma.$$

By Exercise 9.2.6(b), we have

$$\mathscr{S}(\sigma)f(\mathbf{x}) = \begin{cases} A\sigma & \sigma \le \frac{\rho - r}{c} \\ \frac{A}{4cr}[\rho^2 - (r - c\sigma)^2] & \frac{|r-\rho|}{c} \le \sigma \le \frac{r+\rho}{c} \\ 0 & \text{otherwise} \end{cases}$$

where $r = |\mathbf{x}|$. See Figure 22. Therefore

$$u(\mathbf{x}, t) = \int_0^{\min\left\{t, \max\left\{0, \frac{\rho - r}{c}\right\}\right\}} A\sigma\, d\sigma + \int_{\frac{|r-\rho|}{c}}^{\min\left\{\max\left\{t, \frac{|r-\rho|}{c}\right\}, \frac{r+\rho}{c}\right\}} \frac{A}{4cr}\left[\rho^2 - (r - c\sigma)^2\right]\, d\sigma$$

$$= \frac{A}{2}\left[\min\left\{t, \max\left\{0, \frac{\rho - r}{c}\right\}\right\}\right]^2 + \frac{A}{4cr}\left[\rho^2\sigma + \frac{1}{3c}(r - c\sigma)^3\right]\Bigg|_{\frac{|r-\rho|}{c}}^{\min\left\{\max\left\{t, \frac{|r-\rho|}{c}\right\}, \frac{r+\rho}{c}\right\}}.$$

To express the solution in a more reasonable form, we consider six cases, corresponding to the regions in Figure 23.

Case 1: $r > \rho + ct$. In this case

$$\min\left\{t, \max\left\{0, \frac{\rho - r}{c}\right\}\right\} = 0$$

$$\min\left\{\max\left\{t, \frac{|r-\rho|}{c}\right\}, \frac{r+\rho}{c}\right\} = \frac{r - \rho}{c} = \frac{|r - \rho|}{c},$$

126

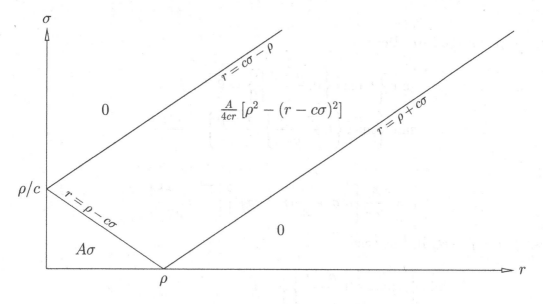

Figure 22: $\mathscr{S}(\sigma)f(\mathbf{x})$

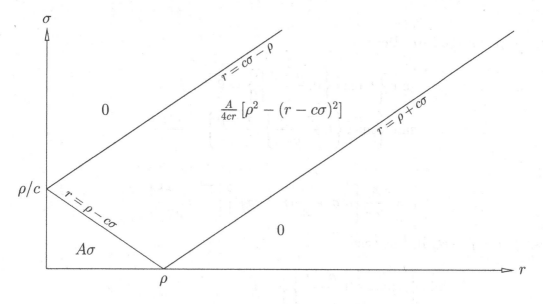

Figure 23: The six cases in Exercise 9.3.7.

so both terms in the solution vanish and therefore $u = 0$.

Case 2: $r > \rho$, $ct - \rho < r < ct + \rho$. In this case,

$$\min\left\{t, \max\left\{0, \frac{\rho - r}{c}\right\}\right\} = 0$$

$$\min\left\{\max\left\{t, \frac{|r - \rho|}{c}\right\}, \frac{r + \rho}{c}\right\} = t,$$

so the first term vanishes and

$$u = \frac{A}{4cr}\left[\rho^2\sigma + \frac{1}{3c}(r - c\sigma)^3\right]\Bigg|_{\frac{r-\rho}{c}}^{t} = \frac{A}{12c^2 r}\left[2\rho^3 + (r - ct)^3 + 3\rho^2(ct - r)\right].$$

127

Case 3: $\rho < r < ct - \rho$. Here,

$$\min\left\{t, \max\left\{0, \frac{\rho - r}{c}\right\}\right\} = 0$$

$$\min\left\{\max\left\{t, \frac{|r - \rho|}{c}\right\}, \frac{r + \rho}{c}\right\} = \frac{r + \rho}{c},$$

so that

$$u = \frac{A}{4cr}\left[\rho^2\sigma + \frac{1}{3c}(r - c\sigma)^3\right]\Big|_{\frac{r-\rho}{c}}^{\frac{r+\rho}{c}} = \frac{\rho^3 A}{3c^2 r}.$$

Case 4: $r < \rho - ct$. In this case

$$\min\left\{t, \max\left\{0, \frac{\rho - r}{c}\right\}\right\} = t$$

$$\min\left\{\max\left\{t, \frac{|r - \rho|}{c}\right\}, \frac{r + \rho}{c}\right\} = \frac{\rho - r}{c} = \frac{|r - \rho|}{c},$$

so the second term in the solution vanishes and $u = \frac{At^2}{2}$.

Case 5: $|ct - \rho| < r < \rho$. In this case,

$$\min\left\{t, \max\left\{0, \frac{\rho - r}{c}\right\}\right\} = \frac{\rho - r}{c}$$

$$\min\left\{\max\left\{t, \frac{|r - \rho|}{c}\right\}, \frac{r + \rho}{c}\right\} = t$$

so that

$$u = \frac{A}{2}\left(\frac{\rho - r}{c}\right)^2 + \frac{A}{4cr}\left[\rho^2\sigma + \frac{1}{3c}(r - c\sigma)^3\right]\Big|_{\frac{\rho-r}{c}}^{t}$$

$$= \frac{A}{12c^2 r}\left[6r(\rho - r)^2 + 3c\rho^2 t + (r - ct)^3 - 3\rho^2(\rho - r) - (2r - \rho)^3\right].$$

Case 6: $r < \rho$, $r < ct - \rho$. In this case,

$$\min\left\{t, \max\left\{0, \frac{\rho - r}{c}\right\}\right\} = \frac{\rho - r}{c}$$

$$\min\left\{\max\left\{t, \frac{|r - \rho|}{c}\right\}, \frac{r + \rho}{c}\right\} = \frac{\rho + r}{c}$$

so that

$$u = \frac{A}{2}\left(\frac{\rho - r}{c}\right)^2 + \frac{A}{4cr}\left[\rho^2\sigma + \frac{1}{3c}(r - c\sigma)^3\right]\Big|_{\frac{\rho-r}{c}}^{\frac{\rho+r}{c}}$$

$$= \frac{A}{6c^2}(3\rho^2 - r^2).$$

Solution 2. We will use equation (9.3.11)

$$u(\mathbf{x}, t) = \int_0^t \mathscr{S}(t - s) f(\mathbf{x}, s) \, ds,$$

where

$$(\mathscr{S}(t)\psi)(\mathbf{x}) = \frac{1}{4\pi c^2 t} \iint_S \psi(\xi) \, d\xi$$

and $S = \{|\xi - \mathbf{x}| = ct\}$. Let $B(\mathbf{x}, t) = \{|\xi - \mathbf{x}| \leq ct\}$. Therefore,

$$u(\mathbf{x}, t) = \int_0^t \frac{1}{4\pi c^2 (t - s)} \iint_{\text{bdy } B(\mathbf{x}, c(t-s))} f(\xi, s) \, dS_\xi \, ds.$$

Since f is zero outside the ball of radius ρ about the origin, and f is constant inside that ball, we can write the solution as

$$u(\mathbf{x}, t) = \int_0^t \frac{1}{4\pi c^2 (t - s)} \iint_{(\text{bdy } B(\mathbf{x}, c(t-s))) \cap B(0, \rho)} A \, dS_\xi \, ds.$$

We note that

$$\iint_{(\text{bdy } B(\mathbf{x}, c(t-s))) \cap B(0, \rho)} dS_\xi$$

is the surface area of that part of the sphere $\{|\xi - \mathbf{x}| \leq c(t - s)\}$ which is contained within the ball of radius ρ about the origin. We will calculate that surface area using spherical coordinates. First, however, we will consider six different cases, corresponding to the six regions in Figure 23.

Case 1: $|\mathbf{x}| - ct > \rho$. In this case (bdy $B(\mathbf{x}, c(t - s))) \cap B(0, \rho) = \emptyset$. Therefore, $u(\mathbf{x}, t) \equiv 0$.

Case 2: $|\mathbf{x}| > \rho$, $|\mathbf{x}| - ct < \rho$, $ct - |\mathbf{x}| < \rho$. For $0 < s < t$, we need to calculate the surface area of that part of the sphere $\{|\xi - \mathbf{x}| \leq c(t - s)\}$ which is contained within the ball of radius ρ about the origin. Now for s such that $|\mathbf{x}| - c(t - s) > \rho$, the intersection of $\{|\xi - \mathbf{x}| \leq c(t - s)\}$ and $|\xi| \leq \rho$ is empty. Therefore, for s such that $|\mathbf{x}| - c(t - s) > \rho$,

$$\iint_{(\text{bdy } B(\mathbf{x}, c(t-s))) \cap B(0, \rho)} dS_\xi = 0.$$

For s such that $|\mathbf{x}| - c(t - s) < \rho$, we can calculate the surface area by using spherical coordinates. In particular, using the law of cosines, we see that

$$\iint_{(\text{bdy } B(\mathbf{x}, c(t-s))) \cap B(0, \rho)} dS_\xi = (c(t - s))^2 \int_0^{2\pi} \int_0^{\theta_0} \sin\theta \, d\theta \, d\phi,$$

where θ_0 satisfies $\rho^2 = |\mathbf{x}|^2 + (c(t-s))^2 - 2|\mathbf{x}|c(t-s)\cos\theta_0$. Therefore,

$$
\iint_{(\text{bdy } B(\mathbf{x},c(t-s)))\cap B(0,\rho)} dS_\xi = (c(t-s))^2 \int_0^{2\pi} \int_0^{\theta_0} \sin\theta \, d\theta \, d\phi
$$

$$
= 2\pi(c(t-s))^2 \left[-\cos\theta\right]\Big|_0^{\theta_0}
$$

$$
= 2\pi(c(t-s))^2[-\cos\theta_0 + 1]
$$

$$
= 2\pi(c(t-s))^2 \left[\frac{\rho^2 - |\mathbf{x}|^2 - (c(t-s))^2}{2|\mathbf{x}|c(t-s)} + \frac{2|\mathbf{x}|c(t-s)}{2|\mathbf{x}|c(t-s)}\right]
$$

$$
= 2\pi(c(t-s))^2 \left[\frac{\rho^2 - (|\mathbf{x}| - c(t-s))^2}{2|\mathbf{x}|c(t-s)}\right]
$$

$$
= \pi(c(t-s)) \left[\frac{\rho^2 - (|\mathbf{x}| - c(t-s))^2}{|\mathbf{x}|}\right].
$$

We put this together to calculate $u(\mathbf{x},t)$. Since there is no intersection if $|\mathbf{x}| - c(t-s) > \rho$, we integrate over s for $0 \le s \le t_0$ where $t_0 = [ct + \rho - |\mathbf{x}|]/c$. Therefore,

$$
u(\mathbf{x},t) = \int_0^{t_0} \frac{1}{4\pi c^2(t-s)} \iint_{\text{bdy } B(\mathbf{x},c(t-s))} A \, dS_\xi \, ds
$$

$$
= \int_0^{t_0} \frac{1}{4\pi c^2(t-s)} A\pi(c(t-s)) \left[\frac{\rho^2 - (|\mathbf{x}| - c(t-s))^2}{|\mathbf{x}|}\right] ds
$$

$$
= \int_0^{t_0} \frac{A}{4c} \left[\frac{\rho^2 - (|\mathbf{x}| - c(t-s))^2}{|\mathbf{x}|}\right] ds
$$

$$
= \frac{A}{4c} \left[\frac{3\rho^2(ct + \rho - |\mathbf{x}|) - \rho^3 + (|\mathbf{x}| - ct)^3}{3c|\mathbf{x}|}\right]
$$

$$
= \frac{A}{12c^2|\mathbf{x}|} \left[2\rho^3 + 3\rho^2(ct - |\mathbf{x}|) + (|\mathbf{x}| - ct)^3\right].
$$

Case 3: $|\mathbf{x}| > \rho$, $ct - |\mathbf{x}| > \rho$. This case is similar to case 2 above, except that now the ball of radius ct about $\mathbf{x}$ contains the ball of radius ρ about the origin. Therefore, the limits of integration with respect to s are now from t_1 to t_2 where t_2 satisfies $|\mathbf{x}| - c(t - t_2) = \rho$ (as t_0 in Case 2 above) and t_1 satisfies $c(t - t_1) - |\mathbf{x}| = \rho$. (For s outside this interval, the intersection of the ball of radius $c(t-s)$ about $\mathbf{x}$ and the ball of radius ρ about the origin is empty.) In this case, our solution is calculated as follows:

$$
u(\mathbf{x},t) = \int_{t_1}^{t_2} \frac{1}{4\pi c^2(t-s)} \iint_{\text{bdy } B(\mathbf{x},c(t-s))} f(\xi,s) \, dS_\xi \, ds
$$

$$
= \int_{t_1}^{t_2} \frac{A}{4c} \left[\frac{\rho^2 - (|\mathbf{x}| - c(t-s))^2}{|\mathbf{x}|}\right] ds
$$

$$
= \frac{A}{4c} \left[\frac{\rho^2 s}{|\mathbf{x}|} - \frac{(|\mathbf{x}| - c(t-s))^3}{3c|\mathbf{x}|}\right]_{s=t_1}^{s=t_2}
$$

where $t_1 = [ct - |\mathbf{x}| - \rho]/c$ and $t_2 = [\rho - |\mathbf{x}| + ct]/c$. This simplifies to

$$
u(\mathbf{x},t) = \frac{A\rho^3}{3c^2|\mathbf{x}|}.
$$

Case 4: $|\mathbf{x}| < \rho$, $|\mathbf{x}| + ct < \rho$. In this case, the ball of radius ct about $\mathbf{x}$ is contained within the ball of radius ρ about the origin. Therefore,

$$u(\mathbf{x}, t) = \int_0^t \frac{1}{4\pi c^2(t - s)} \iint_{\text{bdy } B(x, c(t-s))} A \, dS \, ds$$

$$= \int_0^t A(t - s) \, ds$$

$$= \frac{At^2}{2}.$$

Case 5: $|\mathbf{x}| < \rho$, $|\mathbf{x}| + ct > \rho$, $ct - |\mathbf{x}| < \rho$. Let t_0 be the time such that $|\mathbf{x}| + c(t - t_0) = \rho$. Then

$$u(\mathbf{x}, t) = \int_0^{t_0} \frac{1}{4\pi c^2(t - s)} \iint_{\text{bdy } B(\mathbf{x}, c(t-s))} f(\xi, s) \, dS_\xi \, ds$$

$$+ \int_{t_0}^t \frac{1}{4\pi c^2(t - s)} \iint_{\text{bdy } B(\mathbf{x}, c(t-s))} f(\xi, s) \, dS_\xi \, ds.$$

For $t_0 \leq s \leq t$,

$$\frac{1}{4\pi c^2(t - s)} \iint_{\text{bdy } B(\mathbf{x}, c(t-s))} f(\xi, s) \, dS_\xi = \frac{1}{4\pi c^2(t - s)} \iint_{\text{bdy } B(\mathbf{x}, c(t-s))} A \, dS_\xi$$

$$= A(t - s).$$

Therefore,

$$\int_{t_0}^t \frac{1}{4\pi c^2(t - s)} \iint_{\text{bdy } B(x, c(t-s))} f(\xi, s) \, dS_\xi \, ds = \int_{t_0}^t A(t - s) \, ds$$

$$= \frac{A(t - t_0)^2}{2}.$$

While,

$$\int_0^{t_0} \frac{1}{4\pi c^2(t - s)} \iint_{\text{bdy } B(\mathbf{x}, c(t-s))} f(\xi, s) \, dS_\xi = \frac{A}{4c} \int_0^{t_0} \left[\frac{\rho^2 - (|\mathbf{x}| - c(t - s))^2}{|\mathbf{x}|} \right] ds.$$

Therefore,

$$u(\mathbf{x}, t) = \frac{A(t - t_0)^2}{2} + \frac{A}{4c} \int_0^{t_0} \left[\frac{\rho^2 - (|\mathbf{x}| - c(t - s))^2}{|\mathbf{x}|} \right] ds,$$

where $t_0 = [|\mathbf{x}| + ct - \rho]/c$. This simplifies to

$$u(\mathbf{x}, t) = \frac{A}{4c^2|\mathbf{x}|} \left[\rho^2(|\mathbf{x}| + ct - \rho) - \frac{(2|\mathbf{x}| - \rho)^3}{3} + \frac{(|\mathbf{x}| - ct)^3}{3} + 2|\mathbf{x}|(\rho - |\mathbf{x}|)^2 \right].$$

Case 6: $|\mathbf{x}| < \rho$, $ct - |\mathbf{x}| > \rho$. Let t_1 be the time such that $c(t - t_1) - |\mathbf{x}| = \rho$ and t_2 be the time such that $|\mathbf{x}| + c(t - t_2) = \rho$. Then

$$u(\mathbf{x}, t) = \int_{t_1}^{t_2} \frac{1}{4\pi c^2(t - s)} \iint_{\text{bdy } \mathbf{x}, c(t-s)} f(\xi, s) \, dS_\xi \, ds$$

$$+ \int_{t_2}^t \frac{1}{4\pi c^2(t - s)} \iint_{\text{bdy } B(\mathbf{x}, c(t-s))} f(\xi, s) \, dS_\xi \, ds.$$

131

As in the previous case,

$$\int_{t_1}^{t_2} \frac{1}{4\pi c^2(t-s)} \iint_{\text{bdy } B(\mathbf{x}, c(t-s))} f(\xi, s) \, dS_\xi \, ds = \frac{A}{4c} \int_{t_1}^{t_2} \left[\frac{\rho^2 - (|\mathbf{x}| - c(t-s))^2}{|\mathbf{x}|} \right] ds,$$

while

$$\int_{t_2}^{t} \frac{1}{4\pi c^2(t-s)} \iint_{\text{bdy } B(\mathbf{x}, c(t-s))} f(\xi, s) \, dS_\xi \, ds = \int_{t_2}^{t} A(t-s) \, ds$$

$$= \frac{A(t-t_2)^2}{2}.$$

Therefore,

$$u(\mathbf{x}, t) = \frac{A(t-t_2)^2}{2} + \frac{A}{4c} \int_{t_1}^{t_2} \left[\frac{\rho^2 - (|\mathbf{x}| - c(t-s))^2}{|\mathbf{x}|} \right] ds,$$

where $t_1 = [ct - |\mathbf{x}| - \rho]/c$ and $t_2 = [|\mathbf{x}| + ct - \rho]/c$. This simplifies to

$$u(\mathbf{x}, t) = \frac{A}{6c^2}(3\rho^2 - |\mathbf{x}|^2).$$

9.3.9. We will start using equation (9.3.12),

$$u(\mathbf{x}, t) = \int_0^t \frac{1}{4\pi c^2(t-s)} \iint_{|\xi - \mathbf{x}| = c(t-s)} f(\xi, s) \, dS_\xi \, ds.$$

Next, write the surface integral in spherical coordinates, centered at $\mathbf{x}$, where the angle θ is measured with respect to the line through the origin and $\mathbf{x}$, as shown in Figure 24. We then

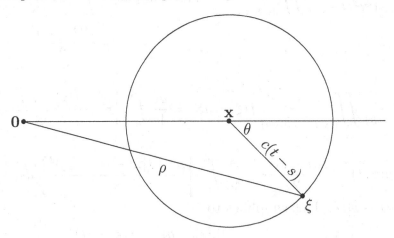

Figure 24: The coordinates in the solution of Exercise 9.3.9.

have

$$\iint_{|\xi - \mathbf{x}| = c(t-s)} f(\xi, s) \, dS_\xi = \int_0^{2\pi} \int_0^\pi f(\xi, s) c^2(t-s)^2 \sin\theta \, d\theta \, d\phi$$

$$= 2\pi c^2(t-s)^2 \int_0^\pi f(\xi, s) \sin\theta \, d\theta.$$

132

Let $\rho = |\xi|$. Then by the Law of Cosines, we have

$$\cos\theta = \frac{|\mathbf{x}|^2 + c^2(t-s)^2 - \rho^2}{2|\mathbf{x}|c(t-s)}$$

so that $\sin\theta\, d\theta = \frac{\rho}{|\mathbf{x}|c(t-s)}\, d\rho$, and thus

$$\int_0^\pi f(\xi,s)\sin\theta\, d\theta = \int_{||\mathbf{x}|-c(t-s)|}^{|\mathbf{x}|+c(t-s)} f(\rho,s)\frac{\rho}{|\mathbf{x}|c(t-s)}\, d\rho.$$

Altogether, this implies

$$u(\mathbf{x},t) = \frac{1}{2c|\mathbf{x}|}\int_0^t \int_{||\mathbf{x}|-c(t-s)|}^{|\mathbf{x}|+c(t-s)} f(\rho,s)\rho\, d\rho\, ds.$$

Section 9.4

9.4.1. If u is a solution of the diffusion equation, then u_{xx} is also a solution. Further, since $u(x,y,z,0) = xy^2z$, we have $u_x(x,y,z,0) = y^2z \implies u_{xx}(x,y,z,0) = 0$. By uniqueness of solutions of the diffusion equation, we have $u_{xx}(x,y,z,t) \equiv 0$. Integrating with respect to x, we see this implies

$$u(x,y,z,t) = A(y,z,t)x + B(y,z,t).$$

We plug a function of this form into the diffusion equation. In particular, $u_t = A_t x + B_t$, $u_{xx} = 0$, $u_{yy} = A_{yy}x + B_{yy}$, $u_{zz} = A_{zz}x + B_{zz}x$ implies

$$A_t x + B_t - k[A_{yy}x + B_{yy}] - k[A_{zz}x + B_{zz}] = 0.$$

We can rewrite this equation as

$$[A_t - kA_{yy} - kA_{zz}]x + [B_t - kB_{yy} - kB_{zz}] = 0.$$

Since this is true for all x, we see A must satisfy the equation

$$A_t - kA_{yy} - kA_{zz} = 0$$

and B must satisfy the equation

$$B_t - kB_{yy} - kB_{zz} = 0.$$

We now use our initial conditions for u to determine what initial conditions A and B must satisfy. Since $u(x,y,z,0) = xy^2z$, we must have

$$u(x,y,z,0) = A(y,z,0)x + B(y,z,0) = xy^2z$$

for all $x \in \mathbb{R}$. Therefore, $B(y, z, 0) = 0$. Since B is a solution of the heat equation in two-dimensions with zero initial conditions, by uniqueness, $B(y, z, t) \equiv 0$. Therefore, $u(x, y, z, t) = A(y, z, t)x$ where A satisfies

$$A_t - kA_{yy} - kA_{zz} = 0$$
$$A(y, z, 0) = y^2 z.$$

We proceed in a similar way as above, now trying to solve for A. If A is a solution of the heat equation, then clearly A_{zz} is also a solution. Further, since $A(y, z, 0) = y^2 z$, $A_{zz}(y, z, 0) = 0$. By uniqueness of solutions to the heat equation, $A_{zz}(y, z, t) \equiv 0$ which implies $A(y, z, t) = C(y, t)z + D(y, t)$. Plugging a function of this form into the equation above for A, we see that

$$C_t z + D_t - kC_{yy}z - kD_{yy} = 0,$$

which we can rewrite as

$$[C_t - kC_{yy}]z + [D_t - kD_{yy}] = 0.$$

Since this is true for all z, we see that

$$C_t - kC_{yy} = 0$$
$$D_t - kD_{yy} = 0.$$

By our initial conditions, $A(y, z, 0) = C(y, 0)z + D(y, 0) = y^2 z$, and thus $C(y, 0) = y^2$ and $D(y, 0) = 0$. By uniqueness of solutions to the heat equation, $D(y, t) \equiv 0$. Therefore, $A(y, z, t) = C(y, t)z$ where C is a solution of

$$C_t - kC_{yy} = 0$$
$$C(y, 0) = y^2.$$

To solve this equation for C, we use the fact that if C is a solution of the heat equation, then C_{yyy} is also a solution. Here, $C_{yyy}(y, 0) = 0$. Therefore, by uniqueness again, $C_{yyy} \equiv 0$. As a result, $C(y, t) = E(t)y^2 + F(t)y + G(t)$. Plugging this function into the heat equation, we see that

$$E_t y^2 + F_t y + G_t - 2kE(t) = 0.$$

By our initial condition, $C(y, 0) = y^2$, we see that

$$E(0)y^2 + F(0)y + G(0) = y^2.$$

Since this is true for all y, we must have $E(0) = 1$ and $F(0) = 0 = G(0)$. Therefore, E, F, G satisfy the following initial-value problems

$$\begin{cases} E_t = 0 \\ E(0) = 1, \end{cases}$$

$$\begin{cases} F_t = 0 \\ F(0) = 0, \end{cases}$$

$$\begin{cases} G_t - 2kE = 0 \\ G(0) = 0. \end{cases}$$

The solutions of these first-order equations are given by $E(t) = 1$, $F(t) = 0$, $G(t) = 2kt$. Therefore,

$$C(y,t) = E(t)y^2 + F(t)y + G(t) = y^2 + 2kt.$$

Working backwards, we see that

$$A(y,z,t) = C(y,t)z = [y^2 + 2kt]z.$$

Finally,

$$u(x,y,z,t) = A(y,z,t)x = [y^2 + 2kt]xz.$$

9.4.3. We reflect the initial data ϕ about the plane $\{z = 0\}$, taking the even reflection. In particular, let

$$\phi_{\text{even}}(x,y,z) = \begin{cases} \phi(x,y,z) & z > 0 \\ \phi(x,y,-z) & z < 0. \end{cases}$$

By equation (9.4.3), our solution is given by

$$\begin{aligned} u(\mathbf{x},t) &= \frac{1}{(4\pi kt)^{3/2}} \int_{-\infty}^{\infty} \int_{-\infty}^{\infty} \int_{-\infty}^{\infty} e^{-\left(\frac{|\mathbf{x}-\mathbf{x}'|^2}{4kt}\right)} \phi_{\text{even}}(\mathbf{x}) \, d\mathbf{x} \\ &= \frac{1}{(4\pi kt)^{3/2}} \int_{0}^{\infty} \int_{-\infty}^{\infty} \int_{-\infty}^{\infty} e^{-\left(\frac{|\mathbf{x}-\mathbf{x}'|^2}{4kt}\right)} \phi(x',y',z') \, dx' \, dy' \, dz' \\ &\quad + \frac{1}{(4\pi kt)^{3/2}} \int_{-\infty}^{0} \int_{-\infty}^{\infty} \int_{-\infty}^{\infty} e^{-\left(\frac{|\mathbf{x}-\mathbf{x}'|^2}{4kt}\right)} \phi(x',y',-z') \, dx' \, dy' \, dz'. \end{aligned}$$

Making the change of variables $\tilde{z} = -z'$ in the last integral, we have

$$\begin{aligned} &\int_{-\infty}^{0} \int_{-\infty}^{\infty} \int_{-\infty}^{\infty} e^{-\left(\frac{|\mathbf{x}-\mathbf{x}'|^2}{4kt}\right)} \phi(x',y',-z') \, dx' \, dy' \, dz' \\ &= -\int_{\infty}^{0} \int_{-\infty}^{\infty} \int_{-\infty}^{\infty} e^{-\left(\frac{(x-x')^2 + (y-y')^2 + (z+\tilde{z})^2}{4kt}\right)} \phi(x',y',\tilde{z}) \, dx' \, dy' \, d\tilde{z} \\ &= \int_{0}^{\infty} \int_{-\infty}^{\infty} \int_{-\infty}^{\infty} e^{-\left(\frac{(x-x')^2 + (y-y')^2 + (z+\tilde{z})^2}{4kt}\right)} \phi(x',y',\tilde{z}) \, dx' \, dy' \, d\tilde{z}. \end{aligned}$$

Therefore,

$$\begin{aligned} u(\mathbf{x},t) &= \frac{1}{(4\pi kt)^{3/2}} \int_{0}^{\infty} \int_{-\infty}^{\infty} \int_{-\infty}^{\infty} \phi(x',y',z') \\ &\quad \cdot \left[e^{-\left(\frac{(x-x')^2 + (y-y')^2 + (z-z')^2}{4kt}\right)} + e^{-\left(\frac{(x-x')^2 + (y-y')^2 + (z+z')^2}{4kt}\right)} \right] dx' \, dy' \, dz'. \end{aligned}$$

9.4.5. We know that the Hermite polynomials are

$$H_k(x) = e^{x^2/2} v_k(x),$$

where v_k is a solution of equation (9.4.15) with $\lambda = 2k + 1$. Let

$$g_k(x) = (-1)^k e^{x^2/2} \frac{\partial^k}{\partial x^k} \left(e^{-x^2} \right).$$

We will show that g_k is a solution of equation (9.4.15) with $\lambda = 2k + 1$ for each k. Indeed,

$$g_k'(x) = (-1)^k x e^{x^2/2} \frac{\partial^k}{\partial x^k} \left(e^{-x^2} \right) + (-1)^k e^{x^2/2} \frac{\partial^{k+1}}{\partial x^{k+1}} \left(e^{-x^2} \right)$$

implies

$$g_k''(x) = (-1)^k e^{x^2/2} \frac{\partial^k}{\partial x^k} \left(e^{-x^2} \right) + (-1)^k x^2 e^{x^2/2} \frac{\partial^k}{\partial x^k} \left(e^{-x^2} \right) + (-1)^k x e^{x^2/2} \frac{\partial^{k+1}}{\partial x^{k+1}} \left(e^{-x^2} \right)$$

$$+ (-1)^k x e^{x^2/2} \frac{\partial^{k+1}}{\partial x^{k+1}} \left(e^{-x^2} \right) + (-1)^k e^{x^2/2} \frac{\partial^{k+2}}{\partial x^{k+2}} \left(e^{-x^2} \right)$$

$$= (-1)^k e^{x^2/2} \left\{ [1 + x^2] \frac{\partial^k}{\partial x^k} \left(e^{-x^2} \right) + [2x] \frac{\partial^{k+1}}{\partial x^{k+1}} \left(e^{-x^2} \right) + \frac{\partial^{k+2}}{\partial x^{k+2}} \left(e^{-x^2} \right) \right\}.$$

Therefore,

$$g_k'' + (2k+1-x^2) g_k = (-1)^k e^{x^2/2} \left\{ [2k+2] \frac{\partial^k}{\partial x^k} \left(e^{-x^2} \right) + [2x] \frac{\partial^{k+1}}{\partial x^{k+1}} \left(e^{-x^2} \right) + \frac{\partial^{k+2}}{\partial x^{k+2}} \left(e^{-x^2} \right) \right\}.$$

We note that

$$\frac{\partial^{k+2}}{\partial x^{k+2}} \left(e^{-x^2} \right) = \frac{\partial^{k+1}}{\partial x^{k+1}} \left(-2x e^{-x^2} \right)$$

$$= -2(k+1) \frac{\partial^k}{\partial x^k} \left(e^{-x^2} \right) - 2x \frac{\partial^{k+1}}{\partial x^{k+1}} \left(e^{-x^2} \right).$$

Therefore, we conclude that

$$g_k'' + (2k + 1 - x^2) g_k = 0$$

for all k.

It is clear that $g_k(x) = e^{-x^2/2} \cdot$(polynomial of degree k), so that $e^{x^2/2} g_k(x)$ is a polynomial of degree k. By the discussion in the text, $e^{x^2/2} g_k(x)$ must be the Hermite polynomial, up to a constant factor.

9.4.6 We know that $H_k(x) = v_k(x) e^{x^2/2}$, where v_k is a solution of $v'' + (\lambda - x^2)v$ for $\lambda = 2k+1$, so that

$$\int_{-\infty}^{\infty} H_k(x) H_l(x) e^{-x^2} \, dx = \int_{-\infty}^{\infty} v_k(x) v_l(x) \, dx.$$

Therefore, it suffices to show that v_k and v_l are orthogonal, where v_k, v_l are eigenfunctions of the eigenvalue problem $v'' + (\lambda - x^2)v$ satisfying the boundary condition $v \to 0$ as $|x| \to \infty$. We do so as follows.

$$\lambda_k \int_{-\infty}^{\infty} v_k v_l \, dx = \int_{-\infty}^{\infty} [x^2 v_k - v_k''] v_l \, dx$$

$$= \int_{-\infty}^{\infty} x^2 v_k v_l \, dx - \int_{-\infty}^{\infty} v_k'' v_l \, dx$$

$$= \int_{-\infty}^{\infty} x^2 v_k v_l \, dx + \int_{-\infty}^{\infty} v_k' v_l' \, dx - v_k' v_l \big|_{-\infty}^{\infty}$$

$$= \int_{-\infty}^{\infty} x^2 v_k v_l \, dx - \int_{-\infty}^{\infty} v_k v_l'' \, dx + v_k v_l' \big|_{-\infty}^{\infty} - v_k' v_l \big|_{-\infty}^{\infty}$$

$$= \int_{-\infty}^{\infty} v_k [x^2 v_l - v_l''] \, dx$$

$$= \int_{-\infty}^{\infty} v_k \lambda_l v_l \, dx$$

$$= \lambda_l \int_{-\infty}^{\infty} v_k v_l \, dx.$$

Therefore,

$$(\lambda_k - \lambda_l) \int_{-\infty}^{\infty} v_k v_l \, dx = 0.$$

Since $\lambda_k = 2k + 1 \neq 2l + 1 = \lambda_l$, we must have

$$\int_{-\infty}^{\infty} v_k v_l \, dx = 0.$$

Therefore, v_k, v_l are orthogonal, and consequently,

$$\int_{-\infty}^{\infty} H_k(x) H_l(x) e^{-x^2} \, dx = 0,$$

as claimed.

Section 9.5

9.5.1. We are looking for solutions of the form $u(\mathbf{x}, t) = v(\mathbf{x}) e^{-i\lambda t/2}$ where v is a solution of

$$-\Delta v - \frac{2}{r} v = \lambda v.$$

Therefore, we just need to verify that the functions v and eigenvalues λ listed in the table in Section 9.5 are solutions of this eigenvalue problem.

Case 1: $\lambda = -1$, $v(\mathbf{x}) = e^{-r}$. Since v depends only on r, we see by equation (6.1.6) that

$$\Delta v = v_{rr} + \frac{2}{r} v_r = \left(-\frac{2}{r} + 1 \right) e^{-r}.$$

137

Therefore,

$$-\Delta v - \frac{2}{r}v = \left(\frac{2}{r} - 1\right)e^{-r} - \frac{2}{r}e^{-r} = -e^{-r} = (-1)v = \lambda v.$$

Case 2: $\lambda = -\frac{1}{4}$, $v(r) = e^{-r/2}\left(1 - \frac{1}{2}r\right)$. Again, since v depends only on r, we have $\Delta v = v_{rr} + \frac{2}{r}v_r$. Since

$$v_r = e^{-r/2}\left(-1 + \frac{1}{4}r\right)$$

$$v_{rr} = e^{-r/2}\left(\frac{3}{4} - \frac{1}{8}r\right)$$

we have

$$\Delta v = e^{-r/2}\left(-\frac{2}{r} + \frac{5}{4} - \frac{r}{8}\right).$$

Therefore,

$$-\Delta v - \frac{2}{r}v = \left[-\frac{1}{4} + \frac{r}{8}\right]e^{-r/2} = -\frac{1}{4}\left[1 - \frac{r}{2}\right]e^{-r/2} = \lambda v.$$

Case 3: $\lambda = -\frac{1}{9}$, $v = e^{-r/3}\left[1 - \frac{2}{3}r + \frac{2}{27}r^2\right]$. Since

$$v_r = e^{-r/3}\left(-1 + \frac{10}{27}r - \frac{2}{81}r^2\right)$$

$$v_{rr} = e^{-r/3}\left(\frac{19}{27} - \frac{14}{81}r + \frac{2}{243}r^2\right)$$

we have

$$\Delta v = v_{rr} + \frac{2}{r}v_r = e^{-r/3}\left(-\frac{2}{r} + \frac{13}{9} - \frac{2}{9}r + \frac{2}{243}r^2\right).$$

Therefore,

$$-\Delta v - \frac{2}{r}v = e^{-r/3}\left(\frac{2}{r} - \frac{13}{9} + \frac{6}{27}r - \frac{2}{243}r^2 - \frac{2}{r} + \frac{4}{3} - \frac{4}{27}r\right)$$

$$= e^{-r/3}\left(-\frac{1}{9} + \frac{2}{27}r - \frac{2}{243}r^2\right)$$

$$= -\frac{1}{9}e^{-r/3}\left(1 - \frac{2}{3}r + \frac{2}{27}r^2\right) = \lambda v.$$

Chapter 10

Section 10.1

10.1.2. Separation of variables in the equation $u_{tt} = c^2(u_{xx} + u_{yy})$ leads to

$$\frac{T''}{c^2 T} = \frac{X''}{X} + \frac{Y''}{Y}.$$

By the usual reasoning, all three quotients are constant, so

$$\frac{X''}{X} = -\lambda, \qquad \frac{Y''}{Y} = -\mu, \qquad \frac{T''}{c^2 T} = -\lambda - \mu$$

for some constants λ and μ. The boundary conditions imply $X(0) = X(a) = Y(0) = Y(b) = 0$, so the ODEs for X and Y have solutions

$$X_m(x) = \sin(m\pi x/a), \qquad Y_n(y) = \sin(m\pi y/b)$$

and the eigenvalues are

$$\lambda_m = \frac{m^2 \pi^2}{a^2}, \qquad \mu_n = \frac{n^2 \pi^2}{b^2}$$

for $m, n \geq 1$. The ODE for T therefore has solution

$$T_{mn}(t) = A_{mn} \cos(c\sqrt{\lambda_m + \mu_n}\, t) + B_{mn} \sin(c\sqrt{\lambda_m + \mu_n}\, t).$$

Now let

$$u(x, y, t) = \sum_{m=1}^{\infty} \sum_{n=1}^{\infty} X_m(x) Y_n(y) \left[A_{mn} \cos(c\sqrt{\lambda_m + \mu_n}\, t) + B_{mn} \sin(c\sqrt{\lambda_m + \mu_n}\, t) \right].$$

The initial condition $u_t(x, y, 0) = 0$ implies $B_{mn} = 0$, and the other initial condition therefore implies

$$xy(a - x)(b - y) = \sum_{m=1}^{\infty} \sum_{n=1}^{\infty} A_{mn} \sin(m\pi x/a) \sin(n\pi y/b).$$

Hence

$$A_{mn} = \frac{4}{ab} \int_0^a \int_0^b xy(a - x)(b - y) \sin(m\pi x/a) \sin(n\pi y/b)\, dy\, dx$$

$$= \left(\frac{2}{a} \int_0^a x(a - x) \sin(m\pi x/a)\, dx \right) \cdot \left(\frac{2}{b} \int_0^b y(b - y) \sin(n\pi y/b)\, dy \right).$$

Integrating by parts in the first integral gives

$$\frac{2}{a}\int_0^a x(a-x)\sin(m\pi x/a)\,dx = \left[\frac{2}{a}\cdot\frac{-a}{m\pi}\cos(m\pi x/a)\right]_0^a - \frac{2}{a}\int_0^a \frac{-a}{m\pi}\cos(m\pi x/a)(a-2x)\,dx$$

$$= \frac{2}{m\pi}\int_0^a \cos(m\pi x/a)(a-2x)\,dx$$

$$= \left[\frac{2}{m\pi}\cdot\frac{a}{m\pi}\sin(m\pi x/a)\right]_0^a - \frac{2}{m\pi}\int_0^a \frac{a}{m\pi}\sin(m\pi x/a)(-2)\,dx$$

$$= \frac{4a}{m^2\pi^2}\int_0^a \sin(m\pi x/a)\,dx$$

$$= \left[-\frac{4a^2}{m^3\pi^3}\cos(m\pi x/a)\right]_0^a$$

$$= \begin{cases} \frac{8a^2}{m^3\pi^3} & m \text{ odd} \\ 0 & m \text{ even.} \end{cases}$$

The second integral is given by the same expression if we replace a with b and m with n. Therefore

$$u(x,y,t) = \sum_{m,n \text{ odd}} \frac{64a^2b^2}{m^3n^3\pi^6}\sin(m\pi x/a)\sin(n\pi y/b)\cos\left(c\pi\sqrt{\frac{m^2}{a^2}+\frac{n^2}{b^2}}\,t\right).$$

10.1.3. Using separation of variables, we have

$$\frac{T'}{T} = k\frac{\Delta X}{X} + \gamma = -\lambda.$$

It then follows that X satisfies the eigenvalue problem

$$\begin{cases} -\Delta X - \frac{\gamma}{k}X = \lambda X & x \in D = (0,a)^3 \\ X = 0 & x \in \text{bdy } D, \end{cases}$$

and $T(t) = e^{-\lambda t}$. Solutions of the PDE $u_t = k\Delta u + \gamma u$ are given by

$$u(\mathbf{x},t) = \sum_{n=1}^\infty X_n(\mathbf{x})e^{-\lambda_n t},$$

where $X_n(\mathbf{x})$ and λ_n are the eigenfunctions and eigenvalues of the problem above. Therefore, in order that the solutions do not grow without bound, we need $\lambda_n \geq 0$ for all n. We know the eigenvalues μ of the eigenvalue problem $-\Delta X = \mu X$ are all positive and are given by

$$\mu = \left(\frac{n_1\pi}{a}\right)^2 + \left(\frac{n_2\pi}{a}\right)^2 + \left(\frac{n_3\pi}{a}\right)^2$$

with $n_1, n_2, n_3 \geq 1$. Therefore, the smallest eigenvalue is

$$\mu_1 = 3\left(\frac{\pi}{a}\right)^2.$$

and the smallest value of λ is

$$\lambda_1 = 3\left(\frac{\pi}{a}\right)^2 - \frac{\gamma}{k}.$$

So $\lambda_n \geq 0$ for all n as long as $\gamma \leq 3k\pi^2/a^2$.

10.1.5.

(a) The solution space of a second order linear homogeneous ODE is two-dimensional because there are two arbitrary constants.

(e) By d'Alembert's formula, any two functions ϕ and ψ determine a solution. Therefore the solution space is infinite dimensional.

Section 10.2

10.2.2. Since the initial data for u is radial and $u_t = 0$ at $t = 0$, the solution takes the form

$$u(r,t) = \sum_{m=1}^{\infty} J_0(\sqrt{\lambda_{0m}}\, r) C_{0m} \cos \sqrt{\lambda_{0m}}\, ct.$$

The initial condition implies

$$1 - \frac{r^2}{a^2} = \sum_{m=1}^{\infty} C_{0m} J_0(\sqrt{\lambda_{0m}}\, r),$$

so the coefficients are given by

$$C_{0m} = \frac{\int_0^a \left(1 - \frac{r^2}{a^2}\right) J_0(\sqrt{\lambda_{0m}}\, r)\, r\, dr}{\int_0^a \left[J_0(\sqrt{\lambda_{0m}}\, r)\right]^2 r\, dr}.$$

10.2.4. In polar coordinates, the wave equation is $u_{tt} = c^2\left(u_{rr} + \frac{1}{r}u_r + \frac{1}{r^2}u_{\theta\theta}\right)$. Inserting $u = e^{-i\omega t} f(r)$ results in the ODE

$$f'' + \frac{1}{r}f' + \frac{\omega^2}{c^2}f = 0.$$

Substituting $f(r) = g(\omega r/c)$ we get $g''(s) + \frac{1}{s}g'(s) + g(s) = 0$. This is Bessel's equation of order zero, so the set of solutions that are finite at the origin is given by $g(s) = CJ_0(s)$. Hence $f(r) = CJ_0(\omega r/c)$ and $u = Ce^{-i\omega t}J_0(\omega r/c)$.

10.2.5. First let $v = u - B$, so $v = 0$ on the boundary and $v = -B$ initially. After separating variables in the equation $v_t = k\Delta v$, the ODE for T is $T' = -\lambda kT$. Thus, assuming a radial solution, it takes the form

$$u(r,t) = B + \sum_{m=1}^{\infty} C_{0m} J_0(\sqrt{\lambda_{0m}}\, r) e^{-k\lambda_{0m} t}.$$

The initial data implies

$$-B = \sum_{m=1}^{\infty} C_{0m} J_0(\sqrt{\lambda_{0m}}r),$$

so the coefficients are

$$C_{0m} = \frac{-B \int_0^a J_0(\sqrt{\lambda_{0m}}r)r\,dr}{\int_0^a \left[J_0(\sqrt{\lambda_{0m}}r)\right]^2 r\,dr}.$$

Both integrals can be evaluated explicitly. By (10.2.22) and (10.5.6),

$$\int_0^a [J_0(\beta r)]^2 r\,dr = \frac{a^2}{2}[J_0'(\beta a)]^2 = \frac{a^2}{2}[J_1(\beta a)]^2,$$

and by Bessel's equation (10.2.10),

$$rJ_0 = -rJ_0'' - J_0' = -(rJ_0')' = -(rJ_1)'.$$

Therefore,

$$\int_0^a rJ_0(r)\,dr = -aJ_1(a)$$

and

$$\int_0^a rJ_0(\beta r)\,dr = -\frac{a}{\beta}J_1(\beta a),$$

so

$$C_{0m} = \frac{2B}{\beta_m a}\frac{1}{J_1(\beta_m a)},$$

where $\beta_m = \sqrt{\lambda_{0m}}$.

10.2.7. Separating variables $u = T(t)R(r)\Theta(\theta)$ in the diffusion equations gives

$$\frac{T'}{kT} = \frac{R''}{R} + \frac{R'}{rR} + \frac{\Theta''}{r^2\Theta}.$$

It then follows that T'/kT must be some constant $-\lambda$, and Θ''/Θ must be a constant $-\gamma$. Thus

$$T' + \lambda kT = 0$$
$$\Theta'' + \gamma\Theta = 0$$
$$R'' + \frac{1}{r}R' + \left(\lambda - \frac{\gamma}{r^2}\right)R = 0.$$

The boundary condition implies that $\Theta(0) = \Theta(\pi) = 0$ and $R(b) = 0$. Thus $\Theta(\theta) = \sin n\theta$ and $\gamma = n^2$, for positive integers n. The equation for R then becomes Bessel's equation after the change of variable $\rho = r\sqrt{\lambda}$. Therefore $R(r) = J_n(\sqrt{\lambda}r)$. Denote by

$$0 < \beta_{n1} < \beta_{n2} < \beta_{n3} < \cdots$$

142

the roots of J_n. Then the boundary condition $R(b) = 0$ is satisfied if $\lambda = \beta_{nm}^2/b^2$ for some m and n. Hence the eigenfunctions of D are $u_{nm}(r, \theta) = \sin(n\theta)J_n(\beta_{nm}r/b)$. The solution of the ODE for T is $T = e^{-k\lambda t}$, so the full solution is

$$u(r, \theta, t) = \sum_{n=1}^{\infty}\sum_{m=1}^{\infty} A_{mn}\sin(n\theta)J_n(\beta_{nm}r/b)e^{-k\beta_{nm}^2 t/b^2},$$

where the coefficients are given by

$$A_{mn} = \frac{2}{\pi j_{nm}}\int_0^b\int_0^\pi \phi(r, \theta)\sin n\theta\, J_n(\beta_{nm}r/b)r\,d\theta\,dr$$

and

$$j_{nm} = \int_0^b [J_n(\beta_{nm}^2 r/b)]^2 r\,dr.$$

Section 10.3

10.3.1. Using equation (10.3.23), the normalizing constants are

$$\int_0^{2\pi}\int_0^\pi |Y_l^m(\theta, \phi)|^2\sin\theta\,d\theta\,d\phi = \int_0^{2\pi}\int_0^\pi [P_l^m(\cos\theta)]^2\sin\theta\,d\theta\,d\phi$$
$$= 2\pi\int_0^\pi [P_l^m(\cos\theta)]^2\sin\theta\,d\theta = \frac{4\pi(l+m)!}{(2l+1)(l-m)!}.$$

10.3.4. For $\lambda = 0$, the only eigenfunctions of $-\Delta$ with boundary condition $u_r = 0$ at $r = a$ are constants, while for positive λ the eigenfunctions are

$$\frac{J_{l+\frac{1}{2}}(\sqrt{\lambda}r)}{\sqrt{r}}P_l^{|m|}(\cos\theta)e^{im\phi},$$

where λ satisfies

$$a\sqrt{\lambda}J'_{l+\frac{1}{2}}(\sqrt{\lambda}a) = \frac{1}{2}J_{l+\frac{1}{2}}(\sqrt{\lambda}a).$$

Denote the roots of this equation by λ_{lj}. Then the complete solution takes the form

$$u = C + \sum_{l=0}^{\infty}\sum_{j=1}^{\infty}\sum_{m=-l}^{l}\frac{J_{l+\frac{1}{2}}(\sqrt{\lambda_{lj}}r)}{\sqrt{r}}P_l^{|m|}(\cos\theta)e^{im\phi}\left[A_{jlm}\cos c\sqrt{\lambda_{lj}}t + B_{jlm}\sin c\sqrt{\lambda_{lj}}t\right].$$

The initial condition $u_t = 0$ implies $B_{jlm} = 0$, while the initial condition $u = r\cos\theta$ implies

$$r\cos\theta = C + \sum_{l=0}^{\infty}\sum_{j=1}^{\infty}\sum_{m=-l}^{l}A_{jlm}\frac{J_{l+\frac{1}{2}}(\sqrt{\lambda_{lj}}r)}{\sqrt{r}}P_l^{|m|}(\cos\theta)e^{im\phi}.$$

143

By the mutual orthogonality of the spherical harmonics $Y_l^m(\theta, \phi) = P_l^{|m|}(\cos\theta)e^{im\phi}$ and the fact that $\cos\theta = Y_1^0(\theta, \phi)$, it follows that $C = 0$ and $A_{jlm} = 0$ for all l and m except $l = 1, m = 0$. Thus

$$r = \sum_{j=1}^{\infty} A_{j10} \frac{J_{\frac{3}{2}}(\sqrt{\lambda_{1j}}r)}{\sqrt{r}},$$

where the coefficients are given by

$$A_{j10} = \frac{\int_0^a J_{\frac{3}{2}}(\sqrt{\lambda_{1j}}r)r^{5/2}\,dr}{\int_0^a [J_{\frac{3}{2}}(\sqrt{\lambda_{1j}}r)]^2 r\,dr}.$$

By equation (10.5.8), the denominator simplifies to

$$\int_0^a [J_{\frac{3}{2}}(\sqrt{\lambda_{1j}}r)]^2 r\,dr = \left(\frac{1}{2}a^2 - \frac{1}{\lambda_{1j}}\right)(J_{3/2}(\sqrt{\lambda_{1j}}a))^2.$$

The solution is therefore

$$u = \sum_{j=1}^{\infty} A_{j10} \frac{J_{3/2}(\sqrt{\lambda_{1j}}\,r)}{\sqrt{r}} \cos\theta \cos(c\sqrt{\lambda_{1j}}t),$$

where

$$a\sqrt{\lambda_{1j}}\,J_{3/2}'(\sqrt{\lambda_{1j}}\,a) = \frac{1}{2}J_{3/2}(\sqrt{\lambda_{1j}}\,a).$$

10.3.5. First let $v = u - B$, so that v satisfies the diffusion equation with homogeneous Dirichlet boundary condition and initial data $v = C - B$ at time $t = 0$. Then since the solution v is radial, only the term $l = 0$, $m = 0$ is present in the expansion given by equation (10.3.22), so

$$v = \sum_{j=1}^{\infty} A_j e^{-k\lambda_{0j}t} \frac{J_{\frac{1}{2}}(\sqrt{\lambda_{0j}}r)}{\sqrt{r}}.$$

Using the expression for $J_{\frac{1}{2}}$ in equation (10.5.11) and the fact that $v(a) = 0$, it follows that $\lambda_{0j} = \frac{j^2\pi^2}{a^2}$ and

$$v = \sum_{j=1}^{\infty} D_j \frac{\sin(j\pi r/a)e^{-\frac{j^2\pi^2 kt}{a^2}}}{r}.$$

The initial condition implies

$$C - B = \sum_{j=1}^{\infty} D_j \frac{\sin(j\pi r/a)}{r},$$

so

$$D_j = \frac{\iiint_{r\leq a}(C-B)\frac{\sin(j\pi r/a)}{r}\,d\mathbf{x}}{\iiint_{r\leq a}\frac{\sin^2(j\pi r/a)}{r^2}\,d\mathbf{x}} = \frac{2a(-1)^{j+1}(C-B)}{\pi j}.$$

144

and

$$u = B + \sum_{j=1}^{\infty} \frac{2a(-1)^{j+1}(C-B)}{\pi j} \cdot \frac{\sin(j\pi r/a)e^{-\frac{j^2\pi^2 kt}{a^2}}}{r}.$$

10.3.6. We may apply the solution of Exercise 10.3.5 with $a = \pi$, $B = 100$ and $C = 20$ to find

$$u = 100 - 160 \sum_{j=1}^{\infty} \frac{(-1)^{j+1}}{j} \cdot \frac{\sin(jr)e^{-0.006j^2 t}}{r}.$$

Using the first term in the expansion at $r = 0$ gives the approximation

$$u(0,t) \approx 100 - 160e^{-0.006t}.$$

Setting this equal to 50 and solving, we get $t = \ln(16/5)/0.006 \approx 193.9$ seconds.

10.3.7.

(a) First make the boundary condition homogeneous. To do so, we look for a solution of $w_t = k(w_{rr} + \frac{2}{r}w_r)$ satisfying $w_r(a) = B$. By guessing a polynomial of the form $w = c_1 t + c_2 r^2$, one finds that $w = \frac{3kBt}{a} + \frac{Br^2}{2a}$ works. Thus if we define $v = u - \frac{3kBt}{a} - \frac{Br^2}{2a}$, it follows that v is a solution of the diffusion equation with boundary data $\partial v/\partial r = 0$ and initial data $v = C - \frac{Br^2}{2a}$ at $t = 0$. As in the solution of Exercise 10.3.4, the eigenfunctions of $-\Delta$ with boundary condition $u_r = 0$ are the constant functions as well as the functions $\frac{J_{l+\frac{1}{2}}(\sqrt{\lambda_{lj}}r)}{\sqrt{r}} P_l^{|m|}(\cos\theta)e^{im\phi}$, where the λ_{lj} are the roots of

$$a\sqrt{\lambda} J'_{l+\frac{1}{2}}(\sqrt{\lambda}a) = \frac{1}{2} J_{l+\frac{1}{2}}(\sqrt{\lambda}a). \tag{S-9}$$

The solution of the diffusion equation therefore has the expansion

$$v(r,\theta,\phi,t) = C_0 + \sum_{l=0}^{\infty} \sum_{j=1}^{\infty} \sum_{m=-l}^{l} A_{jlm} \frac{J_{l+\frac{1}{2}}(\sqrt{\lambda_{lj}}r)}{\sqrt{r}} P_l^{|m|}(\cos\theta)e^{im\phi}e^{-k\lambda_{lj}t}.$$

Inserting the initial condition gives

$$C - \frac{Br^2}{2a} = C_0 + \sum_{l=0}^{\infty} \sum_{j=1}^{\infty} \sum_{m=-l}^{l} A_{jlm} \frac{J_{l+\frac{1}{2}}(\sqrt{\lambda_{lj}}r)}{\sqrt{r}} P_l^{|m|}(\cos\theta)e^{im\phi}, \tag{S-10}$$

so by orthogonality of the constant function 1 to all the other eigenfunctions we have

$$C_0 = \frac{3}{a^3} \int_0^a \left(C - \frac{Br^2}{2a}\right) r^2 \, dr = C - \frac{3}{10}Ba.$$

Since the remaining terms in the expansion of v all decay to zero as $t \to \infty$, the non-decaying terms in the expansion of u are $C - \frac{3}{10}Ba + \frac{3kBt}{a} + \frac{Br^2}{2a}$.

145

(b) To compute the remaining coefficients A_{jlm} first notice that since the left hand side of (S-10) depends only on r, $A_{jlm} = 0$ unless $l = m = 0$. Thus we are left with

$$C - \frac{Br^2}{2a} = C_0 + \sum_{j=1}^{\infty} A_{j00} \frac{J_{\frac{1}{2}}(\sqrt{\lambda_{0j}}r)}{\sqrt{r}}.$$

Since the constant C is orthogonal to each of the non-constant eigenfunctions, the coefficients are given by

$$A_{j00} = \frac{\int_0^a \frac{J_{\frac{1}{2}}(\sqrt{\lambda_{0j}}r)}{\sqrt{r}} \left(\frac{-Br^2}{2a}\right) r^2 \, dr}{\int_0^a [J_{\frac{1}{2}}(\sqrt{\lambda_{0j}}r)]^2 r \, dr}.$$

Using equations (10.5.8) and (10.5.11), the denominator simplifies to

$$\int_0^a [J_{\frac{1}{2}}(\sqrt{\lambda_{0j}}r)]^2 r \, dr = \frac{1}{2}a^2 [J_{1/2}(\sqrt{\lambda_{0j}}a)]^2 = \frac{a\sin^2(\sqrt{\lambda_{0j}}a)}{\pi\sqrt{\lambda_{0j}}}.$$

By equation (10.5.11), equation (S-9) with $l = 0$ becomes $\sin(\sqrt{\lambda}a) = \sqrt{\lambda}a\cos(\sqrt{\lambda}a)$, and the numerator reduces to

$$\int_0^a \frac{J_{\frac{1}{2}}(\sqrt{\lambda_{0j}}r)}{\sqrt{r}} \left(\frac{-Br^2}{2a}\right) r^2 \, dr = -\frac{B}{\sqrt{2\pi}a\lambda_{0j}^{1/4}} \int_0^a \sin(\sqrt{\lambda_{0j}}r)r^3 \, dr$$

$$= -\frac{Ba\sqrt{2}}{\sqrt{\pi}\lambda_{0j}^{5/4}} \sin(a\sqrt{\lambda_{0j}}).$$

So

$$A_{j00} = \frac{-B\sqrt{2\pi}}{\lambda_{0j}^{3/4}\sin(a\sqrt{\lambda_{0j}})}.$$

Therefore the decaying terms in the expansion of u are

$$-2B\sum_{j=1}^{\infty} \frac{\sin(\sqrt{\lambda_{0j}}r)}{\lambda_{0j}r\sin(\sqrt{\lambda_{0j}}a)}e^{-k\lambda_{0j}t}.$$

[Substitution of $\gamma_j = a\sqrt{\lambda_j}$ and $\sin\gamma_j = \gamma_j\cos\gamma_j$ yields the answer in the book.]

10.3.9. For a radial function $u(r,t)$ the diffusion equation takes the form $u_t = k\left(u_{rr} + \frac{2}{r}u_r\right)$. Letting $v = ru$, this becomes $v_t = kv_{rr}$. The boundary conditions are $v(a,t) = 0$ and $v(0,t) = 0$, the latter due to the fact that $v = ru$ and u is finite at $r = 0$. Therefore we may write

$$v(r,t) = \sum_{n=1}^{\infty} A_n \sin(n\pi r/a)e^{-n^2\pi^2 kt/a^2}$$

and hence

$$u(r,t) = \frac{1}{r}\sum_{n=1}^{\infty} A_n \sin(n\pi r/a)e^{-n^2\pi^2 kt/a^2}$$

for some coefficients A_n. The initial condition $v(r,0) = r\phi(r)$ implies that

$$A_n = \frac{2}{a} \int_0^a r\phi(r) \sin(n\pi r/a)\, dr.$$

10.3.10. Separating variables $u = R(r)Y(\theta, \phi)$ leads to equations (10.3.2) and (10.3.3) with $\lambda = 0$. The solution of (10.3.2) is $R(r) = r^\alpha$, where $\alpha^2 + \alpha - \gamma = 0$, and the solution of (10.3.3) is $Y_l^m(\theta, \phi)$ with $\gamma = l(l+1)$. Thus $(\alpha - l)(\alpha + l + 1) = 0$. In order to obtain a solution that is bounded at infinity, we need $\alpha \leq 0$. Thus, when $l = 0$ we may choose $\alpha = 0$ or $\alpha = -1$, but for $l > 0$ we must take $\alpha = -(l+1)$. The series solution therefore takes the form

$$u = A_{01} + A_{02}r^{-1} + \sum_{l=1}^{\infty} \sum_{m=-l}^{l} A_{lm} r^{-(l+1)} P_l^m(\cos\theta)e^{im\phi}.$$

The boundary condition then implies

$$-\cos\theta = -A_{02}a^{-2} + \sum_{l=1}^{\infty} \sum_{m=-l}^{l} -A_{lm}(l+1)a^{-(l+2)} P_l^m(\cos\theta)e^{im\phi}.$$

Now since $Y_1^0 = \cos\theta$ and the spherical harmonics are mutually orthogonal, it follows that $A_{10} = a^3/2$ and $A_{lm} = 0$ for all other l and m, except A_{01}, which is arbitrary. Thus

$$u = A_{01} + \frac{a^3}{2r^2}\cos\theta.$$

10.3.11. First extend the boundary condition to the entire sphere by defining $g(z) = f(z)$ for $z > 0$ and $g(z) = -f(-z)$ for $z < 0$. This defines an odd function of z. Now let u be the harmonic function on the ball with boundary data g. By equation (10.3.25) the solution takes the form

$$u = \sum_{l=0}^{\infty} \sum_{m=-l}^{l} A_{lm} r^l P_l^m(\cos\theta)e^{im\phi}$$

where the coefficients are determined by the equation

$$g(z) = \sum_{l=0}^{\infty} \sum_{m=-l}^{l} A_{lm} a^l P_l^m(\cos\theta)e^{im\phi}.$$

Since $g(z)$ does not depend on ϕ, all coefficients are zero unless $m = 0$, so

$$g(z) = \sum_{l=0}^{\infty} A_{l0} a^l P_l^0(\cos\theta).$$

As stated in the text, the Legendre functions take the form $\sin^m\theta \cdot p(\cos\theta)$, where p is an even polynomial if $l-m$ is even and an odd polynomial if $l-m$ is odd. Thus, for l even and $m = 0$, $g(z)p(\cos\theta) = g(r\cos\theta)p(\cos\theta)$ is an odd function of $\cos\theta$, and therefore the coefficient A_{l0}

is zero. Thus every term in the expansion of the solution u contains a polynomial in $\cos\theta$ of degree at least 1, from which it follows that u vanishes at $\theta = \pi/2$. Therefore the solution u satisfies the boundary condition $u = 0$ on the disk $\{z = 0, x^2 + y^2 < a^2\}$. It now remains to find the coefficients A_{l0} for l odd. They are given by

$$A_{l0} = \frac{2l+1}{2a^l} \cdot \int_0^\pi g(a\cos\theta)P_l^0(\cos\theta)\,d\theta = \frac{2l+1}{a^l} \cdot \int_0^{\pi/2} f(a\cos\theta)P_l^0(\cos\theta)\,d\theta$$

and therefore the final expression for u is

$$u = \sum_{l\ \text{odd}}^\infty (2l+1)\left[\int_0^{\pi/2} f(a\cos\mu)P_l^0(\cos\mu)\,d\mu\right]\left(\frac{r}{a}\right)^l P_l^0(\cos\theta).$$

Section 10.4

10.4.1. The first nine eigenvalues are 2,5,8,10,13,17,18,20 and 25. The eigenvalues 2, 8 and 18 for which $m = n$ each have multiplicity 1, and the rest have multiplicity 2.

10.4.2. Using the identity $\sin 3\theta = \sin\theta(3 - 4\sin^2\theta)$ it follows that $v(x,y) = 2\sin x \sin y(3 - 2\sin^2 x - 2\sin^2 y)$. This vanishes on the boundary of the square and where $\sin^2 x + \sin^2 y = \frac{3}{2}$, and is illustrated in Figure 25.

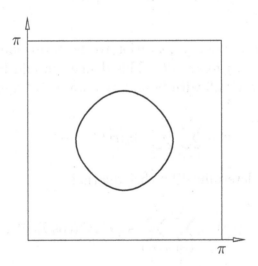

Figure 25: The nodal set of the eigenfunction $\sin 3x \sin y + \sin x \sin 3y$.

Section 10.5

10.5.2. By equation (10.5.11), $J_{1/2}(z) = \sqrt{\frac{2}{\pi z}} \sin z$, so by the recursion formula (10.5.6) with $s = 1/2$,

$$J_{3/2}(z) = \frac{1/2}{z}J_{1/2}(z) - J'_{1/2}(z) = \sqrt{\frac{2}{\pi}}\frac{\sin z - z\cos z}{z^{3/2}}.$$

Since $J_{-1/2}(z) = \sqrt{\frac{2}{\pi z}}\cos z$, equation (10.5.6) with $s = -1/2$ implies

$$J_{-3/2}(z) = -\frac{1/2}{z}J_{-1/2}(z) + J'_{-1/2}(z) = -\sqrt{\frac{2}{\pi}}\frac{\cos z + z\sin z}{z^{3/2}}.$$

10.5.7. Substituting $u = xv$ gives $u' = xv' + v$ and $u'' = xv'' + 2v'$ and thus the ODE becomes $x^2v'' + xv' + (x^2 - 1)v = 0$. Dividing by x^2, this becomes Bessel's equation of order 1. Hence $v(x) = CJ_1(x)$ and $u(x) = CxJ_1(x)$ for some constant C.

10.5.14. Since the boundary condition is radially symmetric, we seek a solution of the form $u(r)$, and the equation takes the form $u_{rr} + \frac{1}{r}u_r = k^2u$. Making the change of variable $u(r) = v(ikr)$ transforms this into $v_{rr} + \frac{1}{r}v_r + v = 0$, which is Bessel's equation of order zero. Thus $v(r) = CJ_0(r)$ for some constant C. Using the boundary condition $u(a) = 1$ it then follows that $u = J_0(ikr)/J_0(ika)$.

10.5.15. As in the solution of Exercise 10.5.14, making the change of variable $u(r) = v(ikr)$ shows that v satisfies Bessel's equation of order zero. Since we want u to be bounded at infinity, we need $v(ikr)$ to be bounded as r goes to infinity. From equation (10.5.15) we see that $H_0^+(ikr)$ is bounded as $r \to \infty$. So applying the boundary condition $u(a) = 1$, we have $u(r) = H_0^+(ikr)/H_0^+(ika)$.

10.5.16. Since the boundary condition is radial, we may assume a radial solution, in which case the equation takes the form $u_{rr} + \frac{2}{r}u_r = k^2u$. Making the change of variable $u(r) = r^{-1/2}v(ikr)$ leads to the equation $v_{zz} + \frac{1}{z}v_z + \left(1 - \frac{1/4}{z^2}\right)v = 0$, which is Bessel's equation of order 1/2. Therefore $v(z) = CJ_{1/2}(z) = C\sqrt{\frac{2}{\pi z}}\sin z$ for some constant C, which implies $u(r) = Ar^{-1}\sinh(kr)$ for some constant A. The boundary condition $u(a) = 1$ then implies $u = \frac{a\sinh kr}{r\sinh ka}$.

Another approach is to set $v = ru$. This leads to the equation $v_{rr} = k^2v$, so $v = c_1\cosh kr + c_2\sinh kr$. Since u is bounded and equals 1 at $r = a$, v must satisfy the boundary conditions $v(0) = 0$ and $v(a) = a$. Thus $v = a\sinh kr/\sinh ka$ and $u = \frac{a\sinh kr}{r\sinh ka}$.

10.5.18. In polar coordinates, we have $-u_{rr} - \frac{1}{r}u_r - \frac{1}{r^2}u_{\theta\theta} = \lambda u$ so separation of variables leads to the equations $\Theta'' = -\gamma\Theta$ and $R'' + \frac{1}{r}R' + \left(\lambda - \frac{\gamma}{r^2}\right)R = 0$. The periodic boundary conditions in θ imply that $\Theta(\theta) = A_n\cos n\theta + B_n\sin n\theta$, and $\gamma = n^2$. Thus, after the change of variable $\rho = \sqrt{\lambda}r$, $R(\rho)$ satisfies Bessel's equation of order n. Thus, in order for u to be non-singular at $r = 0$, $R(\rho) = J_n(\rho)$ and therefore $u(r, \theta) = J_n(\sqrt{\lambda}r)(A_n\cos n\theta + B_n\sin n\theta)$. In order to satisfy the boundary condition $u_r + hu = 0$ at $r = a$, we need λ to be a root of $\sqrt{\lambda}J'_n(\sqrt{\lambda}a) + hJ_n(\sqrt{\lambda}a) = 0$.

10.5.19. As shown in the solution to Exercise 10.5.18, $\Theta = A_n\cos n\theta + B_n\sin n\theta$ and $R(\rho)$ is a solution of Bessel's equation of order n, where $\rho = \sqrt{\lambda}r$. The general solution of Bessel's equation is $R(\rho) = C_n J_n(\rho) + D_n N_n(\rho)$. Hence

$$u(r, \theta) = (C_n J_n(\sqrt{\lambda}r) + D_n N_n(\sqrt{\lambda}r))(A_n \cos n\theta + B_n \sin n\theta).$$

The boundary condition $u(a, \theta) = 0$ implies $D_n = -\frac{J_n(\sqrt{\lambda}a)}{N_n(\sqrt{\lambda}a)}C_n$, so after absorbing the remaining constant into A_n and B_n, we have

$$u(r, \theta) = (N_n(\sqrt{\lambda}a)J_n(\sqrt{\lambda}r) - J_n(\sqrt{\lambda}a)N_n(\sqrt{\lambda}r))(A_n \cos n\theta + B_n \sin n\theta).$$

The boundary condition $u(b, \theta) = 0$ then implies that λ must be a root of

$$N_n(\sqrt{\lambda}a)J_n(\sqrt{\lambda}b) - J_n(\sqrt{\lambda}a)N_n(\sqrt{\lambda}b) = 0.$$

Section 10.6

10.6.5. For any solution u of Legendre's equation, integration by parts gives

$$\gamma \int_0^1 xu(x)\, dx = -\int_0^1 x[(1-x^2)u'(x)]'\, dx$$

$$= -x(1-x^2)u'(x)\Big|_0^1 + \int_0^1 (1-x^2)u'(x)\, dx$$

$$= u(x)(1-x^2)\Big|_0^1 + 2\int_0^1 xu(x)\, dx,$$

so

$$\int_0^1 xu(x)\, dx = -\frac{u(0)}{\gamma - 2}$$

for $\gamma \neq 2$. Applying this with $\gamma = l(l+1)$ and $u = P_l$ gives

$$\int_{-1}^1 f(x)P_l(x)\, dx = \int_0^1 xP_l(x)\, dx = -\frac{P_l(0)}{(l-1)(l+2)},$$

for $l \neq 1$. By equation (10.6.3), $P_l(0) = 0$ for all odd l and

$$P_{2n}(0) = \frac{(-1)^n(2n)!}{2^{2n}(n!)^2}.$$

Hence,

$$\int_{-1}^1 f(x)P_{2n}(x)\, dx = \frac{(-1)^{n+1}(2n-2)!}{2^{2n}(n+1)!(n-1)!}$$

for $n \geq 1$. Combining this with equation (10.6.6) gives

$$a_{2n} = \frac{(-1)^{n+1}(4n+1)(2n-2)!}{2 \cdot 4^n(n+1)!(n-1)!}$$

150

for $n \geq 1$. The indices $l = 0$ and $l = 1$ are special cases. Since $P_0(0) = 1$, it follows that $a_0 = \frac{1}{4}$. For $l = 1$,

$$\int_{-1}^{1} f(x) P_1(x)\, dx = \int_{0}^{1} x^2\, dx = \frac{1}{3},$$

so $a_1 = \frac{1}{2}$.

10.6.6. The solution of Laplace's equation is given in terms of Legendre polynomials in $\cos\theta$ by equation (10.3.25):

$$u = \sum_{l=0}^{\infty} \sum_{m=-l}^{l} A_{lm} r^l P_l^m(\cos\theta) e^{im\phi}.$$

The boundary condition then implies

$$\cos^2\theta = \sum_{l=0}^{\infty} \sum_{m=-l}^{l} A_{lm} a^l P_l^m(\cos\theta) e^{im\phi}.$$

Since $P_0^0(z) = 1$ and $P_2^0(z) = \frac{1}{2}(3z^2 - 1)$, it is easy to see that $z^2 = \frac{2}{3} P_2^0(z) + \frac{1}{3} P_0^0(z)$. Then by the orthogonality of the eigenfunctions, it follows that $A_{00} = \frac{1}{3}$, $A_{20} = \frac{2}{3a^2}$, and $A_{lm} = 0$ for all other l and m. So $u = \frac{1}{3} + \frac{2}{3}(r^2/a^2) P_2^0(\cos\theta) = \frac{1}{3} + \frac{r^2}{a^2}\left(\cos^2\theta - \frac{1}{3}\right)$.

10.6.7. By equation (10.3.25),

$$u = \sum_{l=0}^{\infty} \sum_{m=-l}^{l} A_{lm} r^l P_l^m(\cos\theta) e^{im\phi}.$$

The boundary condition then implies

$$f(\theta) = \sum_{l=0}^{\infty} \sum_{m=-l}^{l} A_{lm} a^l P_l^m(\cos\theta) e^{im\phi},$$

where $f(\theta) = A$ for $0 \leq \theta < \pi/2$ and $f(\theta) = B$ for $\pi/2 < \theta \leq \pi$. Since the left hand side is independent of ϕ, this becomes

$$f(\theta) = \sum_{l=0}^{\infty} A_{l0} a^l P_l(\cos\theta),$$

and therefore the coefficients are

$$A_{l0} = \frac{2l+1}{2a^l} \int_{0}^{\pi} f(\theta) P_l(\cos\theta) \sin\theta\, d\theta.$$

For $l = 0$ this gives $A_{00} = \frac{1}{2}(A + B)$, while for $l \neq 0$, we have

$$\int_{0}^{\pi} f(\theta) P_l(\cos\theta) \sin\theta\, d\theta = B \int_{0}^{\pi} P_l(\cos\theta) \sin\theta\, d\theta + (A - B) \int_{0}^{\pi/2} P_l(\cos\theta) \sin\theta\, d\theta$$

$$= (A - B) \int_{0}^{\pi/2} P_l(\cos\theta) \sin\theta\, d\theta.$$

Making the change of variable $s = \cos\theta$ gives

$$\int_0^{\pi/2} P_l(\cos\theta)\sin\theta\, d\theta = \int_0^1 P_l(s)\, ds.$$

Using equation (10.6.1) this becomes

$$-\frac{1}{l(l+1)}\int_0^1 [(1-s^2)P_l'(s)]'\, ds = \frac{1}{l(l+1)}P_l'(0).$$

By equation (10.6.3), $P_l'(0) = 0$ for all even l, and

$$P_{2n+1}'(0) = \frac{(-1)^n(2n+2)!}{n!(n+1)!2^{2n+1}}.$$

for odd $l = 2n+1$. Therefore $A_{l0} = 0$ for even $l \geq 2$ and for odd $l = 2n+1$,

$$A_{l0} = \frac{(A-B)(-1)^n(2n+2)!(4n+3)}{n!(n+1)!2^{2n+1}2a^l(2n+1)(2n+2)} = \frac{(A-B)(-1)^n(2n)!(4n+3)}{n!(n+1)!2^{2n+2}a^{2n+1}}.$$

The solution is therefore

$$u = \frac{1}{2}(A+B) + \sum_{n=0}^{\infty} \frac{(A-B)(-1)^n(2n)!(4n+3)}{n!(n+1)!2^{2n+2}}\left(\frac{r}{a}\right)^{2n+1} P_{2n+1}(\cos\theta).$$

10.6.8 Begin with $u_t = k\Delta u$ in the cone $D = \{r < a, 0 \leq \theta < \alpha\}$. Separate variables exactly as in Section 10.3. Thus a separated solution is

$$u = e^{-\lambda kt}R(r)P(\theta)q(\phi)$$

where R satisfies (10.3.2), P satisfies (10.3.12), and q satisfies (10.3.11). Since $R(0)$ is finite, we have (10.3.8). By its periodicity in ϕ, q has the form written below (10.3.11).

The difference from the text is that now the boundary conditions for $P(\theta)$ are

$$P \text{ finite at } \theta = 0, \quad P(\alpha) = 0.$$

Converting to $s = \cos\theta$ as in the text, and writing $p(s) = P(\cos\theta)$, we get the associated Legendre equation (10.3.14)

$$[(1-s^2)p_s]_s + \left(\gamma - \frac{m^2}{1-s^2}\right)p = 0$$

in the interval $\cos\alpha < s < 1$ and

$$p(1) \text{ finite}, \quad p(\cos\alpha) = 0.$$

[It is no longer true that $\gamma = l(l+1)$.]

152

Notice that $s = 1$ is a regular singular point of this ODE (see Section A.4). It is therefore most natural to expand the solution in powers of $(s - 1)$. We follow Section A.4 to look for a series expansion of the form (A.4.6):

$$p(s) = C(s - 1)^{\sigma_1} \sum_{n=0}^{\infty} p_n(s - 1)^n + D(s - 1)^{\sigma_2} \sum_{n=0}^{\infty} q_n(s - 1)^n$$

or possibly a similar form with a logarithm as in (A.4.7). We should keep in mind that m is a nonnegative integer. Let us find the indicial equation (A.4.5). We rewrite the ODE as

$$p_{ss} - \frac{2s}{1 - s^2} p_s + \left[\frac{\gamma}{1 - s^2} - \frac{m^2}{(1 - s^2)^2} \right] p = 0$$

so that the indicial equation has the coefficients

$$\beta = \lim_{s \to 1}(s - 1)\frac{-2s}{1 - s^2} = \lim_{s \to 1}\frac{2s}{s + 1} = 1$$

and

$$\delta = \lim_{s \to 1}(s - 1)^2 \left[\frac{\gamma}{1 - s^2} - \frac{m^2}{(1 - s^2)^2} \right] = \lim_{s \to 1}\left[\frac{\gamma(1 - s)}{1 + s} - \frac{m^2}{(1 + s)^2} \right] = -\frac{m^2}{4}.$$

Therefore the indicial equation is $\sigma(\sigma - 1) + \beta\sigma + \delta = 0$ or $\sigma^2 = m^2/4$. It has the roots $\sigma_1 = -m/2$ and $\sigma_2 = +m/2$. Therefore

$$p(s) = C(s - 1)^{-m/2} \sum_{n=0}^{\infty} p_n(s - 1)^n + [Cb \log(s - 1) + D](s - 1)^{+m/2} \sum_{n=0}^{\infty} q_n(s - 1)^n$$

for some fixed constant b and for arbitrary constants C and D.

If $m \neq 0$, the boundary condition that $p(1)$ is finite requires $C = 0$. We may take $D = 1$ for simplicity. Thus

$$p(s) = (s - 1)^{+m/2} \sum_{n=0}^{\infty} q_n(s - 1)^n. \tag{S-11}$$

It is of course possible to determine the coefficients q_n recursively.

As described in Section A.4, there is the exceptional case when $m = 0$, in which case $\sigma_1 = \sigma_2 = 0$ and

$$p(s) = C \sum_{n=0}^{\infty} p_n(s - 1)^n + [Cb \log(s - 1) + D] \sum_{n=0}^{\infty} q_n(s - 1)^n.$$

The boundary condition requires $b = 0$, so that

$$p(s) = \sum_{n=0}^{\infty} q_n(s - 1)^n \quad \text{if } m = 0.$$

Thus (S-11) is the correct formula for all m.

We want to find the equation for the eigenvalues. Let's now denote the solution (S-11) for $p(s)$ by $p_{\gamma m}(s)$. The other boundary condition requires

$$p_{\gamma m}(\cos \alpha) = 0.$$

This requirement determines a sequence $\gamma_{m\ell}$ for $\ell = 1, 2, \ldots$ in terms of m and the constant α. Summing up, we have the general solution

$$u = \sum_{m\ell j} A_{m\ell j} e^{-\lambda_{m\ell j} kt} \, e^{im\phi} \frac{1}{\sqrt{r}} \, J_{\sqrt{\gamma_{m\ell}+\frac{1}{4}}}(\sqrt{\lambda_{m\ell j}} r) \; p_{\gamma_m \ell m}(\cos \theta).$$

The eigenvalue equation is obtained by putting in the boundary condition at the outer surface $r = a$ of the cone; namely, $\lambda = \lambda_{m\ell j}$ satisfies the equation

$$J_{\sqrt{\gamma_{m\ell}+\frac{1}{4}}}(\sqrt{\lambda} a) = 0.$$

That is, $\sqrt{\lambda} a$ is a zero of a certain Bessel function. The sequence of zeroes is numbered $j = 1, 2, \ldots$.

Section 10.7

10.7.2. Using the definition in equation (10.7.5),

$$\begin{aligned}
L_x L_y - L_y L_x &= (-i)^2 (y\partial_z - z\partial_y)(z\partial_x - x\partial_z) - (-i)^2 (z\partial_x - x\partial_z)(y\partial_z - z\partial_y) \\
&= -(y\partial_x + yz\partial_{xz} - yx\partial_{zz} - z^2\partial_{xy} + zx\partial_{zy}) \\
&\quad + (zy\partial_{zx} - z^2\partial_{yx} - xy\partial_{zz} + xz\partial_{yz} + x\partial_y) \\
&= x\partial_y - y\partial_x = iL_z,
\end{aligned}$$

$$\begin{aligned}
L_y L_z - L_z L_y &= (-i)^2 (z\partial_x - x\partial_z)(x\partial_y - y\partial_x) - (-i)^2 (x\partial_y - y\partial_x)(z\partial_x - x\partial_z) \\
&= -(z\partial_y + zx\partial_{yx} - zy\partial_{xx} - x^2\partial_{yz} + xy\partial_{xz}) \\
&\quad + (xz\partial_{xy} - x^2\partial_{zy} - yz\partial_{xx} + yx\partial_{zx} + y\partial_z) \\
&= y\partial_z - z\partial_y = iL_x,
\end{aligned}$$

and

$$\begin{aligned}
L_z L_x - L_x L_z &= (-i)^2 (x\partial_y - y\partial_x)(y\partial_z - z\partial_y) - (-i)^2 (y\partial_z - z\partial_y)(x\partial_y - y\partial_x) \\
&= -(x\partial_z + xy\partial_{zy} - xz\partial_{yy} - y^2\partial_{zx} + yz\partial_{yx}) \\
&\quad + (yx\partial_{yz} - y^2\partial_{xz} - zx\partial_{yy} + zy\partial_{xy} + z\partial_x) \\
&= z\partial_x - x\partial_z = iL_y,
\end{aligned}$$

10.7.3.

(a) For $n = 1$, we must have $l = m = 0$, so by equation (10.7.16), the solution is $v_{1,0,0} = e^{-r}L_1^0(r)Y_0^0(r) = e^{-r}$.

(b) For $n = 2$, we must have $l = 0$, in which case $m = 0$, or $l = 1$, in which case $m = -1, 0, 1$. Thus the solutions are

$$v_{2,0,0} = e^{-r/2}L_2^0(r)Y_0^0(\theta, \phi)$$
$$v_{2,1,0} = e^{-r/2}L_2^1(r)Y_1^0(\theta, \phi)$$
$$v_{2,1,1} = e^{-r/2}L_2^1(r)Y_1^1(\theta, \phi)$$
$$v_{2,1,-1} = e^{-r/2}L_2^1(r)Y_1^{-1}(\theta, \phi).$$

Using the recursion relation (10.7.14), we have $L_2^0(r) = 1 - \frac{1}{2}r$ and $L_2^1(r) = r$. Using the equations in Section 10.3 that define the spherical harmonics, it follows that

$$v_{2,0,0} = e^{-r/2}\left(1 - \frac{1}{2}r\right)$$
$$v_{2,1,0} = e^{-r/2}r\cos\theta$$
$$v_{2,1,1} = e^{-r/2}r\sin\theta e^{i\phi}$$
$$v_{2,1,-1} = e^{-r/2}r\sin\theta e^{-i\phi}.$$

(c) For $n = 3$, the eigenfunctions are

$$v_{3,0,0} = e^{-r/3}L_3^0(r)Y_0^0(\theta, \phi)$$
$$v_{3,1,0} = e^{-r/3}L_3^1(r)Y_1^0(\theta, \phi)$$
$$v_{3,1,1} = e^{-r/3}L_3^1(r)Y_1^1(\theta, \phi)$$
$$v_{3,1,-1} = e^{-r/3}L_3^1(r)Y_1^{-1}(\theta, \phi)$$
$$v_{3,2,0} = e^{-r/3}L_3^2(r)Y_2^0(\theta, \phi)$$
$$v_{3,2,1} = e^{-r/3}L_3^2(r)Y_2^1(\theta, \phi)$$
$$v_{3,2,-1} = e^{-r/3}L_3^2(r)Y_2^{-1}(\theta, \phi)$$
$$v_{3,2,2} = e^{-r/3}L_3^2(r)Y_2^2(\theta, \phi)$$
$$v_{3,2,-2} = e^{-r/3}L_3^2(r)Y_2^{-2}(\theta, \phi).$$

Using the recursion equation (10.7.14) gives $L_3^0(r) = 1 - \frac{2}{3}r + \frac{2}{27}r^2$, $L_3^1(r) = r - \frac{1}{6}r^2$, and

$L_3^2(r) = r^2$. Hence

$$v_{3,0,0} = e^{-r/3}\left(1 - \frac{2}{3}r + \frac{2}{27}r^2\right)$$

$$v_{3,1,0} = e^{-r/3}\left(r - \frac{1}{6}r^2\right)\cos\theta$$

$$v_{3,1,1} = e^{-r/3}\left(r - \frac{1}{6}r^2\right)\sin\theta e^{i\phi}$$

$$v_{3,1,-1} = e^{-r/3}\left(r - \frac{1}{6}r^2\right)\sin\theta e^{-i\phi}$$

$$v_{3,2,0} = e^{-r/3}r^2\left(3\cos^2\theta - 1\right)$$

$$v_{3,2,1} = e^{-r/3}r^2\sin\theta\cos\theta e^{i\phi}$$

$$v_{3,2,-1} = e^{-r/3}r^2\sin\theta\cos\theta e^{-i\phi}$$

$$v_{3,2,2} = e^{-r/3}r^2\sin^2\theta e^{2i\phi}$$

$$v_{3,2,-2} = e^{-r/3}r^2\sin^2\theta e^{-2i\phi}.$$

10.7.5.

(a) Inserting the separated solution $u = T(t)R(r)\Theta(\theta)$ gives

$$\frac{2iT'}{T} = -\left(\frac{R''}{R} + \frac{1}{r}\frac{R'}{R} + \frac{1}{r^2}\frac{\Theta''}{\Theta}\right) + 2V(r)$$

so that $T' = -\frac{1}{2}i\lambda T$ for some constant λ, $\Theta'' = -\gamma\Theta$ for some constant γ. The periodic boundary condition $\Theta(0) = \Theta(2\pi)$ implies $\gamma = n^2$, so the equation for R is

$$R'' + \frac{1}{r}R' + \left(\lambda - \frac{n^2}{r^2} - 2V(r)\right)R = 0.$$

(b) With $V(r) = \frac{1}{2}r^2$, we have

$$R_{rr} + \frac{1}{r}R_r + \left[-r^2 + \lambda - \frac{n^2}{r^2}\right]R = 0.$$

First substitute $R(r) = r^{-n}w(r)$. Thus

$$R_r = r^{-n}w_r - nr^{-n-1}w$$

$$R_{rr} = r^{-n}w_{rr} - 2nr^{-n-1}w_r + n(n+1)r^{-n-2}w$$

and the differential equation becomes

$$w_{rr} - \frac{2n}{r}w_r + \frac{n(n+1)}{r^2}w + \frac{1}{r}w_r - \frac{n}{r^2}w + \left[-r^2 + \lambda - \frac{n^2}{r^2}\right]w = 0,$$

or equivalently

$$w_{rr} + \frac{1-2n}{r}w_r + (-r^2 + \lambda)w = 0.$$

Next, we substitute $\rho = r^2$, $w_r = 2\sqrt{\rho}w_\rho$, $w_{rr} = 4\rho w_{\rho\rho} + 2w_\rho$ to obtain

$$4\rho w_{\rho\rho} + 2w_\rho + \frac{1-2n}{\sqrt{\rho}}(2\sqrt{\rho}w_\rho) + (-\rho + \lambda)w = 0,$$

or equivalently

$$w_{\rho\rho} + \frac{1-n}{\rho}w_\rho + \left(-\frac{1}{4} + \frac{\lambda}{4\rho}\right)w = 0.$$

For very large ρ, this equation looks like $w_{\rho\rho} = \frac{1}{4}w$. Next, we substitute $w(r) = e^{-\rho/2}L(\rho)$. The combined transformation is

$$R(r) = r^{-n}w(r) = \rho^{-n/2}e^{-\rho/2}L(\rho).$$

The substitution $w = e^{-\rho/2}L(\rho)$ gives us

$$w_\rho = e^{-\rho/2}\left(L_\rho - \frac{1}{2}L\right)$$

$$w_{\rho\rho} = e^{-\rho/2}\left(L_{\rho\rho} - L_\rho + \frac{1}{4}L\right)$$

so that

$$\left(L_{\rho\rho} - L_\rho + \frac{1}{4}L\right) + \frac{1-n}{\rho}\left(L_\rho - \frac{1}{2}L\right) + \left(-\frac{1}{4} + \frac{\lambda}{4\rho}\right)L = 0.$$

Thus

$$L_{\rho\rho} + \left(-1 + \frac{1-n}{\rho}\right)L_\rho + \left(\frac{n-1}{2} + \frac{\lambda}{4}\right)\frac{1}{\rho}L = 0,$$

which is the Laguerre equation with $\nu = -n$ and $\mu = \frac{n-1}{2} + \frac{\lambda}{4}$.

Chapter 11

Section 11.1

11.1.1. The first eigenfunction of $-u_{xx} = \lambda u$ on $[0,3]$ is $\sin(\pi x/3)$, so the first eigenvalue is $\lambda_1 = \pi^2/9 > 1$. If there were a function f such that $\int_0^3 [f'(x)]\, dx = \int_0^3 [f(x)]\, dx = 1$ and $f(0) = f(3) = 0$, then by Theorem 11.1.1,

$$\lambda_1 \le \frac{\int_0^3 [f'(x)]^2\, dx}{\int_0^3 [f(x)]^2} = 1,$$

a contradiction. Therefore there is no such f.

11.1.2. Let

$$m_1 = \min\left\{ \frac{\|\nabla w\|^2}{\|w\|^2} : w \in C^2, w = 0 \text{ on bdy } D, w \not\equiv 0 \right\}$$

and let

$$m_2 = \min\left\{ \|\nabla v\|^2 : v \in C^2, v = 0 \text{ on bdy } D, \|v\|^2 = 1 \right\}.$$

We need to show $m_1 = m_2$. Let w be any C^2 function such that $w = 0$ on bdy D and $w \not\equiv 0$, and let

$$m_1 = \frac{\|\nabla w\|^2}{\|w\|^2}.$$

Then if we define $v = w/\|w\|$, v is C^2 and satisfies $v = 0$ on bdy D and $\|v\|^2 = 1$. Therefore

$$m_2 \le \|\nabla v\|^2 = \frac{\|\nabla w\|^2}{\|w\|^2} = m_1.$$

On the other hand, suppose v is a C^2 function such that $v = 0$ on bdy D, $\|v\|^2 = 1$ and

$$m_2 = \|\nabla v\|^2.$$

Then $v \not\equiv 0$ and therefore

$$m_1 \le \frac{\|\nabla v\|^2}{\|v\|^2} = \|\nabla v\|^2 = m_2.$$

11.1.5.

(a) Suppose the function u minimizes the quotient. Then

$$m = \frac{\iiint_D |\nabla u|^2\, d\mathbf{x} + \iint_{\text{bdy } D} a|u|^2\, dS}{\iiint_D |u|^2\, d\mathbf{x}} \le \frac{\iiint_D |\nabla w|^2\, d\mathbf{x} + \iint_{\text{bdy } D} a|w|^2\, dS}{\iiint_D |w|^2\, d\mathbf{x}}$$

for all C^2 functions w such that $w \not\equiv 0$. Let $w = u + \epsilon v$, where v is any C^2 function. Then the function

$$f(\epsilon) = \frac{\iiint_D |\nabla(u + \epsilon v)|^2\, d\mathbf{x} + \iint_{\text{bdy } D} a|u + \epsilon v|^2\, dS}{\iiint_D |u + \epsilon v|^2\, d\mathbf{x}}$$

158

has a local minimum at $\epsilon = 0$, so $f'(0) = 0$. Expanding the terms in the integrals gives

$$f(\epsilon) = \frac{\iiint_D |\nabla u|^2 + 2\epsilon \nabla u \cdot \nabla v + \epsilon^2 |\nabla v|^2 \, d\mathbf{x} + \iint_{\text{bdy } D} a(|u|^2 + 2\epsilon uv + \epsilon^2 |v|^2) \, dS}{\iiint_D |u|^2 + 2\epsilon uv + \epsilon^2 |v|^2 \, d\mathbf{x}},$$

so

$$0 = f'(0) = \frac{\|u\|^2 \left[2\int_D \nabla u \cdot \nabla v + 2 \int_{\text{bdy } D} auv \right] - \left[\int_D |\nabla u|^2 + \int_{\text{bdy } D} a|u|^2 \right] 2 \int_D uv}{\|u\|^4}$$

and thus

$$\frac{\iiint_D \nabla u \cdot \nabla v \, d\mathbf{x} + \iint_{\text{bdy } D} auv \, dS}{\iiint_D uv \, d\mathbf{x}} = \frac{\iiint_D |\nabla u|^2 \, d\mathbf{x} + \iint_{\text{bdy } D} a|u|^2 \, dS}{\iiint_D |u|^2 \, d\mathbf{x}} = m.$$

This implies

$$\iiint_D \nabla u \cdot \nabla v - muv \, d\mathbf{x} + \iint_{\text{bdy } D} auv \, dS = 0$$

and by Green's first identity,

$$\iiint_D (-\Delta u - mu)v \, d\mathbf{x} + \iint_{\text{bdy } D} \left(\frac{\partial u}{\partial n} + au \right) v \, dS = 0.$$

This is true for any nonzero C^2 function v. In particular, for all nonzero C^2 functions v which vanish on bdy D we have

$$\iiint_D (-\Delta u - mu)v \, d\mathbf{x} = 0.$$

and thus $-\Delta u = mu$. The last two equations now imply

$$\iint_{\text{bdy } D} \left(\frac{\partial u}{\partial n} + au \right) v \, dS = 0.$$

Letting v be any function which equals $\frac{\partial u}{\partial n} + au$ on bdy D, it then follows that

$$\iint_{\text{bdy } D} \left| \frac{\partial u}{\partial n} + au \right|^2 dS = 0$$

and thus $\frac{\partial u}{\partial n} + au = 0$ on bdy D. Hence u is an eigenfunction with eigenvalue m. To show that m is the smallest eigenvalue, suppose v is an eigenfunction with any eigenvalue λ. Then, applying Green's first identity gives

$$m \leq \frac{\iiint_D |\nabla v|^2 \, d\mathbf{x} + \iint_{\text{bdy } D} a|v|^2 \, dS}{\iiint_D |v|^2 \, d\mathbf{x}}$$

$$= \frac{\iiint_D -v\Delta v \, d\mathbf{x} + \iint_{\text{bdy } D} \left(\frac{\partial v}{\partial n} + av \right) v \, dS}{\iiint_D |v|^2 \, d\mathbf{x}}$$

$$= \frac{\lambda \iiint_D |v|^2 \, d\mathbf{x}}{\iiint_D |v|^2 \, d\mathbf{x}}$$

$$= \lambda,$$

so m is the smallest eigenvalue.

(b) Suppose $a_1(\mathbf{x}) \leq a_2(\mathbf{x})$ for all $\mathbf{x} \in D$, and let $\lambda_1(a_1)$ and $\lambda_1(a_2)$ denote the first eigenvalues of $-\Delta$ with boundary conditions $\frac{\partial u}{\partial n} + a_1 u = 0$ and $\frac{\partial u}{\partial n} + a_1 u = 0$, respectively. Then $\lambda_1(a_1) = \min I_1(w)$ and $\lambda_1(a_2) = \min I_2(w)$, where

$$I_1(w) = \frac{\iiint_D |\nabla w|^2\, d\mathbf{x} + \iint_{\text{bdy } D} a_1(\mathbf{x})|w|^2\, dS}{\iiint_D |w|^2\, d\mathbf{x}},$$

$$I_2(w) = \frac{\iiint_D |\nabla w|^2\, d\mathbf{x} + \iint_{\text{bdy } D} a_2(\mathbf{x})|w|^2\, dS}{\iiint_D |w|^2\, d\mathbf{x}},$$

and the minimum is taken over all nonzero C^2 functions w. Now let u_2 be the minimizer of I_2. Then

$$\lambda_1(a_1) = \min I_1(w) \leq I_1(u_2) \leq I_2(u_2) = \min I_2(w) = \lambda_1(a_2),$$

so λ_1 is increasing in a.

Section 11.2

11.2.1.

$$a_{11} = \int_0^1 (1-2x)^2\, dx = \int_0^1 1 - 4x + 4x^2\, dx = 1 - 2 + \frac{4}{3} = \frac{1}{3}$$

$$a_{12} = a_{21} = \int_0^1 (1-2x)(2x-3x^2)\, dx = \int_0^1 2x - 7x^2 + 6x^3\, dx = 1 - \frac{7}{3} + \frac{3}{2} = \frac{1}{6}$$

$$a_{22} = \int_0^1 (2x-3x^2)^2\, dx = \int_0^1 4x^2 - 12x^3 + 9x^4\, dx = \frac{4}{3} - 3 + \frac{9}{5} = \frac{2}{15}$$

$$b_{11} = \int_0^1 (x-x^2)^2\, dx = \int_0^1 x^2 - 2x^3 + x^4\, dx = \frac{1}{3} - \frac{1}{2} + \frac{1}{5} = \frac{1}{30}$$

$$b_{12} = b_{21} = \int_0^1 (x-x^2)(x^2-x^3)\, dx = \int_0^1 x^3 - 2x^4 + x^5\, dx = \frac{1}{4} - \frac{2}{5} + \frac{1}{6} = \frac{1}{60}$$

$$b_{22} = \int_0^1 (x^2-x^3)^2\, dx = \int_0^1 x^4 - 2x^5 + x^6\, dx = \frac{1}{5} - \frac{1}{3} + \frac{1}{7} = \frac{1}{105}$$

So

$$\det(A - \lambda B) = \det \begin{bmatrix} \frac{1}{3} - \frac{1}{30}\lambda & \frac{1}{6} - \frac{1}{60}\lambda \\ \frac{1}{6} - \frac{1}{60}\lambda & \frac{2}{15} - \frac{1}{105}\lambda \end{bmatrix}$$

$$= \left(\frac{1}{3} - \frac{1}{30}\lambda\right)\left(\frac{2}{15} - \frac{1}{105}\lambda\right) - \left(\frac{1}{6} - \frac{1}{60}\lambda\right)^2$$

$$= \frac{1}{25200}\lambda^2 - \frac{13}{6300}\lambda + \frac{1}{60}$$

$$= \frac{(\lambda - 10)(\lambda - 42)}{25200}$$

160

and thus $\lambda_1 \approx 10$ and $\lambda_2 \approx 42$. The actual eigenvalues are $\lambda_1 = \pi^2 \approx 9.8696$ and $\lambda_2 = 4\pi^2 \approx 39.4784$.

11.2.3. Write $w(x,y) = xy(\pi - x)(\pi - y)$. Then $\nabla w(x,y) = (y(\pi - y)(\pi - 2x), x(\pi - x)(\pi - 2y))$, so

$$\|\nabla w\|^2 = \int_0^\pi \int_0^\pi y^2(\pi - y)^2(\pi - 2x)^2 + x^2(\pi - x)^2(\pi - 2y)^2 \, dy \, dx$$

$$= 2 \int_0^\pi y^2(\pi - y)^2 \, dy \int_0^\pi (\pi - 2x)^2 \, dx$$

$$= 2 \cdot \frac{\pi^5}{30} \cdot \frac{\pi^3}{3} = \frac{1}{45}\pi^8.$$

Since

$$\|w\|^2 = \int_0^\pi \int_0^\pi x^2 y^2 (\pi - x)^2 (\pi - y)^2 \, dy \, dx$$

$$= \int_0^\pi x^2(\pi - x)^2 \, dx \int_0^\pi y^2(\pi - y)^2 \, dy$$

$$= \frac{\pi^5}{30} \cdot \frac{\pi^5}{30} = \frac{\pi^{10}}{900}$$

the Rayleigh quotient for w is

$$\frac{\|\nabla w\|^2}{\|w\|^2} = \frac{20}{\pi^2} \approx 2.0264.$$

The first eigenvalue is $\lambda_1 = 1^2 + 1^2 = 2$.

11.2.4. It can be shown by separation of variables that the *first* eigenfunction is radial. Now recall that if u is a radial solution of $-\Delta u = \lambda u$, then substituting $v = ru$ gives $v_{rr} = -\lambda v$. Hence

$$v(r) = A\cos(\sqrt{\lambda}r) + B\sin(\sqrt{\lambda}r).$$

Since $v(0) = 0u(0) = 0$, we have $A = 0$. The boundary condition $v(1) = 1u(1) = 0$ then implies that $\sqrt{\lambda} = n\pi$ for some positive integer n. Hence the first eigenvalue is $\lambda_1 = \pi^2$.

(a) Let $w = 1 - r$. Then $\nabla w = (-x/r, -y/r, -z/r)$, so

$$\|\nabla w\|^2 = \iiint_{r<1} 1 \, d\mathbf{x} = \frac{4}{3}\pi.$$

Since

$$\|w\|^2 = \iiint_{r<1} (1 - r)^2 \, d\mathbf{x} = \int_0^{2\pi} \int_0^\pi \int_0^1 (1 - r)^2 r^2 \sin\theta \, dr \, d\theta \, d\phi = \frac{2}{15}\pi,$$

the Rayleigh quotient for w is

$$\frac{\|\nabla w\|^2}{\|w\|^2} = 10.$$

The actual value of λ_1 is $\pi^2 \approx 9.8696$.

161

(b) Let $w = \cos \frac{1}{2}\pi r$. Then $\nabla w = -\frac{1}{2}\pi \sin \frac{1}{2}\pi r(-x/r, -y/r, -z/r)$, so

$$\|\nabla w\|^2 = \iiint_{r<1} \frac{\pi^2}{4} \sin^2 \frac{1}{2}\pi r \, d\mathbf{x}$$

$$= \frac{\pi^2}{4} \int_0^{2\pi} \int_0^{\pi} \int_0^1 r^2 \sin^2 \frac{1}{2}\pi r \sin\theta \, dr \, d\theta \, d\phi$$

$$= \pi^3 \int_0^1 r^2 \left(\frac{1}{2} - \frac{1}{2}\cos\pi r \right) dr$$

$$= \frac{\pi^3}{2} \left[\int_0^1 r^2 \, dr - \int_0^1 r^2 \cos\pi r \, dr \right]$$

$$= \frac{\pi^3}{2} \left[\frac{1}{3} + \frac{2}{\pi^2} \right].$$

Likewise,

$$\|w\|^2 = \iiint_{r<1} \cos^2 \frac{1}{2}\pi r \, d\mathbf{x}$$

$$= \int_0^{2\pi} \int_0^{\pi} \int_0^1 r^2 \cos^2 \frac{1}{2}\pi r \sin\theta \, dr \, d\theta \, d\phi$$

$$= 4\pi \int_0^1 r^2 \left(\frac{1}{2} + \frac{1}{2}\cos\pi r \right) dr$$

$$= 2\pi \left[\int_0^1 r^2 \, dr + \int_0^1 r^2 \cos\pi r \, dr \right]$$

$$= 2\pi \left[\frac{1}{3} - \frac{2}{\pi^2} \right].$$

Thus the Rayleigh quotient for w is

$$\frac{\|\nabla w\|}{\|w\|} = \frac{\pi^2(\pi^2 + 6)}{4(\pi^2 - 6)} \approx 10.119.$$

The actual value of λ_1 is $\pi^2 \approx 9.8696$.

11.2.6

(a) $A = (a_{jk})$ where $a_{jk} = (\nabla w_j, \nabla w_k) = \int_D \nabla w_j \cdot \nabla w_k \, dx$. Here $w_1 = 1 - x^2$, $w_2 = 1 - x^3$, $D = (0, 1)$, $\nabla = d/dx$. Therefore,

$$a_{11} = \int_0^1 w_1' \cdot w_1' \, dx = \int_0^1 (-2x)^2 \, dx = 4 \int_0^1 x^2 \, dx = \frac{4}{3},$$

$$a_{12} = a_{21} = \int_0^1 w_1' \cdot w_2' \, dx = \int_0^1 (-2x)(-3x^2) \, dx = 6 \int_0^1 x^3 \, dx = \frac{3}{2}.$$

162

and

$$a_{22} = \int_0^1 w_2' \cdot w_2' \, dx = \int_0^1 (-3x^2)(-3x^2) \, dx = 9 \int_0^1 x^4 \, dx = \frac{9}{5}.$$

Therefore,

$$A = \begin{bmatrix} 4/3 & 3/2 \\ 3/2 & 9/5 \end{bmatrix}.$$

Next, $B = (b_{jk})$ where $b_{jk} = \int_D w_j w_j \, dx$. Therefore,

$$b_{11} = \int_0^1 (1 - x^2)^2 \, dx = \int_0^1 (1 - 2x^2 + x^4) \, dx = x - \frac{2}{3}x^3 + \frac{x^5}{5}\Big|_0^1 = \frac{8}{15},$$

$$b_{12} = b_{21} = \int_0^1 (1 - x^2)(1 - x^3) \, dx = \int_0^1 (1 - x^2 - x^3 + x^5) \, dx = x - \frac{x^3}{3} - \frac{x^4}{4} + \frac{x^6}{6}\Big|_0^1 = \frac{7}{12},$$

and

$$b_{22} = \int_0^1 (1 - x^3)^2 \, dx = \int_0^1 (1 - 2x^3 + x^6) \, dx = x - \frac{x^4}{2} + \frac{x^7}{7}\Big|_0^1 = \frac{9}{14}.$$

Therefore,

$$B = \begin{bmatrix} 8/15 & 7/12 \\ 7/12 & 9/14 \end{bmatrix}.$$

(b) Using the results of part (a),

$$\det(A - \lambda B) = \det \begin{bmatrix} \frac{4}{3} - \frac{8}{15}\lambda & \frac{3}{2} - \frac{7}{12}\lambda \\ \frac{3}{2} - \frac{7}{12}\lambda & \frac{9}{5} - \frac{9}{14}\lambda \end{bmatrix}$$

$$= \left(\frac{4}{3} - \frac{8}{15}\lambda\right)\left(\frac{9}{5} - \frac{9}{14}\lambda\right) - \left(\frac{3}{2} - \frac{7}{12}\lambda\right)^2$$

$$= \frac{13}{5040}\lambda^2 - \frac{141}{2100}\lambda + \frac{3}{20}.$$

Therefore, $65\lambda^2 - 1692\lambda + 3780 = 0$, so

$$\lambda = \frac{1692 \pm \sqrt{1880064}}{130} \approx 23.6, 2.47.$$

(c) Now we will find the exact eigenvalues. We will solve

$$\begin{cases} -u'' = \lambda u & 0 < x < 1 \\ u'(0) = u(1) = 0. \end{cases}$$

It is straightforward to verify that all eigenvalues are positive. If $\lambda = \beta^2 > 0$, then $u(x) = A\cos(\beta x) + B\sin(\beta x)$. Therefore, $u(1) = A\cos(\beta) + B\sin(\beta) = 0$. Also, $u'(x) = -\beta A\sin(\beta x) + \beta B\cos(\beta x)$ implies $u'(0) = B\beta$. Therefore, $u'(0) = 0$ implies $B = 0$. Then $u(1) = A\cos(\beta) = 0$ implies $\beta = n\pi - \pi/2$ for $n = 1, 2, \ldots$ Therefore, the first eigenvalue is $\lambda_1 = \beta_1^2 = (\pi/2)^2 = \pi^2/4 \approx 2.47$. The second eigenvalue is $\lambda_2 = \beta_2^2 = (3\pi/2)^2 = 9\pi^2/4 \approx 22.2$.

11.2.8.

(a) We will frequently use the formula

$$\int_0^a t^n (a-t)^m \, dt = \frac{n!\,m!\,a^{n+m+1}}{(n+m+1)!}$$

which holds for all nonnegative integers m and n and all real $a > 0$. Let $w(x,y) = xy(1-x-y)$. Then

$$\begin{aligned}
\|w\|^2 &= \int_0^1 \int_0^{1-x} x^2 y^2 (1-x-y)^2 \, dy \, dx \\
&= \int_0^1 x^2 \int_0^{1-x} y^2 (1-x-y)^2 \, dy \, dx \\
&= \frac{2!\,2!}{5!} \int_0^1 x^2 (1-x)^5 \, dx \\
&= \frac{2!\,2!}{5!} \frac{2!\,5!}{8!} = \frac{1}{5040}.
\end{aligned}$$

Next, $\nabla w(x,y) = (y(1-y-2x), x(1-x-2y))$, so

$$\begin{aligned}
\|\nabla w\|^2 &= \iint_D y^2 (1-y-2x)^2 + x^2 (1-x-2y)^2 \, dA \\
&= 2 \int_0^1 \int_0^{1-y} y^2 (1-y-2x)^2 \, dx \, dy \\
&= 2 \int_0^1 y^2 \left[\frac{(1-y-2x)^3}{-6} \right]_0^{1-y} dy \\
&= \frac{2}{3} \int_0^1 y^2 (1-y)^3 \, dy \\
&= \frac{2}{3} \cdot \frac{2!\,3!}{6!} = \frac{1}{90}.
\end{aligned}$$

Thus the Rayleigh quotient for w is

$$\frac{\|\nabla w\|^2}{\|w\|^2} = 56.$$

(b) Let $w_1(x,y) = xy(1-x-y)$ and $w_2(x,y) = [w_1(x,y)]^2$. From part (a), $a_{11} = \frac{1}{90}$ and $b_{11} = \frac{1}{30} \cdot \frac{1}{168}$. Since $\nabla w_2 = 2w_1 \nabla w_1$, we have

$$\begin{aligned}
a_{22} &= \iint_D 4x^2 y^2 (1-x-y)^2 [y^2(1-y-2x)^2 + x^2(1-x-2y)^2] \, dA \\
&= 8 \int_0^1 y^4 \int_0^{1-y} x^2 (1-x-y)^2 (1-y-2x)^2 \, dx \, dy.
\end{aligned}$$

Integrating by parts several times, the inner integral equals

$$
\left[\frac{(1-y-2x)^3}{-6}x^2(1-x-y)^2\right]_0^{1-y} + \frac{1}{3}\int_0^{1-y} x(1-x-y)(1-y-2x)^4\, dy
$$

$$
= \left[\frac{(1-y-2x)^5}{-30}x(1-x-y)\right]_0^{1-y} + \frac{1}{30}\int_0^{1-y}(1-y-2x)^6\, dx
$$

$$
= \left[\frac{(1-y-2x)^7}{-420}\right]_0^{1-y} = \frac{(1-y)^7}{210}.
$$

Thus

$$
a_{22} = \frac{8}{210}\int_0^1 y^4(1-y)^7\, dy = \frac{8}{210}\frac{4!7!}{12!}.
$$

Next,

$$
a_{12} = a_{21} = \iint_D 2xy(1-x-y)[y^2(1-y-2x)^2 + x^2(1-x-2y)^2]\, dA
$$

$$
= 4\int_0^1 y^3 \int_0^{1-y} x(1-x-y)(1-y-2x)^2\, dx\, dy
$$

$$
= 4\int_0^1 y^3 \int_0^{1-y} \frac{(1-y-2x)^4}{6}\, dx\, dy
$$

$$
= \frac{2}{15}\int_0^1 y^3(1-y)^5\, dy = \frac{2}{15}\frac{3!5!}{9!}.
$$

Next,

$$
b_{22} = \iint_D x^4 y^4 (1-x-y)^4\, dA
$$

$$
= \int_0^1 y^4 \int_0^{1-y} x^4(1-x-y)^4\, dx\, dy
$$

$$
= \frac{4!4!}{9!}\int_0^1 y^4(1-y)^9\, dy = \frac{4!4!}{9!}\frac{4!9!}{14!}
$$

Finally,

$$
b_{12} = b_{21} = \iint_D x^3 y^3 (1-x-y)^3\, dA
$$

$$
= \int_0^1 y^3 \int_0^{1-y} x^3(1-x-y)^3\, dx\, dy
$$

$$
= \frac{3!3!}{7!}\int_0^1 y^3(1-y)^7\, dy = \frac{3!3!}{7!}\frac{3!7!}{11!}.
$$

Using MATLAB, the roots of $\det(A - \lambda B)$ are approximately 53.397 and 316.875.

Section 11.3

11.3.1. We know that the eigenvalues of the Neumann problem are nonnegative. Thus it suffices to show that $\tilde{\lambda}_2 \neq 0$. This amounts to showing that there are not two linearly independent eigenfunctions with eigenvalue zero. Clearly any nonzero constant function is an eigenfunction with eigenvalue zero. We claim that these are the only eigenfunctions with zero eigenvalue. For suppose v satisfies $-\Delta v = 0$ on D and $\frac{\partial v}{\partial n} = 0$ on bdy D. Then by Green's first identity,

$$0 = \iiint_D -v\Delta v \, d\mathbf{x} = \iint_{\text{bdy } D} -v\frac{\partial v}{\partial n} \, dS + \iiint_D |\nabla v|^2 \, d\mathbf{x} = \iiint_D |\nabla v|^2 \, d\mathbf{x},$$

and thus by the First Vanishing Theorem, $\nabla v = 0$ on D and thus v is constant. Hence the eigenspace with eigenvalue zero is one-dimensional, so $\tilde{\lambda}_2 > 0$.

11.3.2. Suppose u minimizes the given functional among all C^2 functions w. Then for any C^2 function v,

$$F(\epsilon) = \frac{1}{2}\iiint_D |\nabla u + \epsilon \nabla v|^2 \, d\mathbf{x} - \iiint_D (u + \epsilon v)f \, d\mathbf{x} - \iint_{\text{bdy } D} (u + \epsilon v)g \, dS$$

has a local minimum at $\epsilon = 0$, so

$$0 = F'(0) = \iiint_D \nabla u \cdot \nabla v \, d\mathbf{x} - \iiint_D fv \, d\mathbf{x} - \iint_{\text{bdy } D} gv \, dS.$$

By Green's first identity, this becomes

$$\iiint_D (-\Delta u - f)v \, d\mathbf{x} + \iint_{\text{bdy } D} \left(\frac{\partial u}{\partial n} - g\right) v \, dS = 0.$$

Since this holds for all C^2 functions v, it follows that

$$\iiint_D (-\Delta u - f)v \, d\mathbf{x} = 0.$$

for all C^2 functions v which vanish on bdy D. It then follows in the usual way that $-\Delta u = f$ on D, and thus

$$\iint_{\text{bdy } D} \left(\frac{\partial u}{\partial n} - g\right) v \, dS = 0.$$

Finally, by choosing v to equal $\frac{\partial u}{\partial n} - g$ on bdy D, it follows that $\frac{\partial u}{\partial n} = g$ on bdy D.

Section 11.4

11.4.1.

(a) Let $x = e^s$. Then

$$-(xu')' = \frac{\lambda}{x}u \implies -(e^s u_x)_x = \frac{\lambda}{e^s}u.$$

Now if $x = e^s$, then for a given function u, by the chain rule, $u_x = u_s/x_s = u_s/e^s$. Therefore,

$$-\left(e^s \frac{u_s}{e^s}\right)_x = \frac{\lambda}{e^s}u \implies -(u_s)_x = \frac{\lambda}{e^s}u$$

$$\implies \frac{-u_{ss}}{x_s} = \frac{\lambda}{e^s}u$$

$$\implies \frac{-u_{ss}}{e^s} = \frac{\lambda}{e^s}u$$

$$\implies -u_{ss} = \lambda u.$$

With this change of variables $x = e^s$, our new interval becomes $0 < s < \log(b)$. Therefore, we need to solve the eigenvalue problem

$$\begin{cases} -u_{ss} = \lambda u & 0 < s < \log(b) \\ u(0) = 0 = u(\log(b)). \end{cases}$$

We know the solutions of this eigenvalue problem are given by $u_n(s) = \sin(n\pi s/\log(b))$, $\lambda_n = (n\pi/\log(b))^2$. Making a change of variables back, we see our solutions are given by $u_n(x) = \sin(n\pi \log(x)/\log(b))$ with $\lambda_n = (n\pi/\log(b))^2$.

(b) By equation (11.4.4), the inner product is defined as

$$(f, g) = \int_1^b \frac{1}{x} f(x)\overline{g(x)}\, dx.$$

Therefore, for this eigenvalue problem, we say two functions are orthogonal if $(f, g) = 0$. That is, if

$$\int_1^b \frac{1}{x} f(x)\overline{g(x)}\, dx = 0.$$

11.4.4 We look for a solution of the form $u(x, y) = X(x)Y(y)$. Plugging such a function into our PDE, we have

$$(x^2 X')'Y + x^2 XY'' = 0.$$

Dividing this equation by $x^2 XY$, we have

$$\frac{(x^2 X')'}{x^2 X} + \frac{Y''}{Y} = 0,$$

which implies

$$\frac{(x^2 X')'}{x^2 X} = -\frac{Y''}{Y} = -\lambda.$$

First, we need to solve

$$\begin{cases} (x^2 X')' = -\lambda x^2 X & 1 < x < 2 \\ X(1) = 0 = X(2). \end{cases}$$

Thus

$$xX'' + 2X' + \lambda x X = 0.$$

Now we notice that $xX'' + 2X' = (xX)''$. Therefore, our eigenvalue problem can be written as

$$(xX)'' + \lambda x X = 0.$$

For simplicity, introduce the function $v(x) = xX$. Therefore, our eigenvalue problem reduces to

$$v'' + \lambda v = 0$$

with the boundary conditions $v(1) = 0 = v(2)$. We know the eigenvalues for this problem are $\lambda_n = (n\pi)^2$ with corresponding eigenfunctions $v_n(x) = \sin(n\pi x)$. Therefore, $X_n(x) = \sin(n\pi x)/x$. Now we need to solve our equation for Y. Our problem is

$$-Y_n'' = -\lambda_n Y_n \qquad -1 < y < 1.$$

Since $\lambda_n = (n\pi)^2 > 0$, we know these solutions are given by $Y_n(y) = C_n \cosh(n\pi y) + D_n \sinh(n\pi y)$. Therefore,

$$u(x,y) = \sum_{n=1}^{\infty} \frac{\sin(n\pi x)}{x} \left[C_n \cosh(n\pi y) + D_n \sinh(n\pi y) \right].$$

Now we need $u(x,1) = f(x) = u(x,-1)$. Therefore, we look for coefficients C_n and D_n such that

$$f(x) = \sum_{n=1}^{\infty} \frac{\sin(n\pi x)}{x} \left[C_n \cosh(n\pi) + D_n \sinh(n\pi) \right]$$

$$= \sum_{n=1}^{\infty} \frac{\sin(n\pi x)}{x} \left[C_n \cosh(n\pi) - D_n \sinh(n\pi) \right].$$

It then follows that $D_n = 0$ and

$$C_n = \frac{2}{\cosh(n\pi)} \int_1^2 x f(x) \sin(n\pi x)\, dx.$$

11.4.5 We use the method of shifting the data of Section 5.6. Let $v = u - \frac{J_0(\omega x)}{J_0(\omega l)} \cos(\omega t)$,

168

where J_0 is the Bessel function of order zero. Then v satisfies

$$\begin{cases} v_{tt} - \left(\dfrac{1}{x}\right)(xv_x)_x = 0 \\[2mm] |v(0,t)| < \infty \\[2mm] v(l,t) = 0 \\[2mm] v(x,0) = \phi(x) - \dfrac{J_0(\omega x)}{J_0(\omega l)} \\[2mm] v_t(x,0) = \psi(x). \end{cases}$$

We look for a solution of the form $v(x,t) = X(x)T(t)$. Plugging a function of this form into our equation, we have

$$XT'' = \frac{1}{x}(xX')'T$$

$$\implies \frac{XT''}{XT} = \frac{(xX')'T}{xXT}$$

$$\implies \frac{T''}{T} = \frac{(xX')'}{xX} = -\lambda.$$

Now

$$T'' = -\lambda T \implies T(t) = A_n \cos(\sqrt{\lambda_n}t) + B_n \sin(\sqrt{\lambda_n}t).$$

Next, we need to solve

$$xX'' + X' + \lambda x X = 0.$$

Dividing this equation by x, we arrive at the equation

$$X'' + \frac{1}{x}X' + \lambda X = 0.$$

We saw this ODE in Section 10.2. (See equation (10.2.8).) The solutions of this equation are given by the Bessel function of order 0, $X(x) = J_0(\sqrt{\lambda}x)$. Therefore, $v(x,t) = XT = J_0(\sqrt{\lambda}x)[A_n \cos(\sqrt{\lambda_n}t) + B_n \sin(\sqrt{\lambda_n}t)]$. Now we need to satisfy the boundary conditions. First, the condition $|v(0)| < \infty$ is already satisfied because $J_0(0)$ is bounded. Second, we need $v(l,t) = 0$. Now $v(l,t) = J_0(\sqrt{\lambda_n}l)[A_n \cos(\sqrt{\lambda_n}t) + B_n \sin(\sqrt{\lambda_n}t)] = 0$ for all t implies $J_0(\sqrt{\lambda_n}l) = 0$. In other words, the eigenvalues λ_n are roots of $J_0(\sqrt{\lambda_n}l) = 0$. Therefore,

$$v(x,t) = \sum_{n=1}^{\infty} J_0(\sqrt{\lambda_n}x)[A_n \cos(\sqrt{\lambda_n}t) + B_n \sin(\sqrt{\lambda_n}t)],$$

where λ_n are the solutions of $J_0(\sqrt{\lambda_n}l) = 0$. In order for the initial conditions to be satisfied, we need

$$v(x,0) = \sum_{n=1}^{\infty} A_n J_0(\sqrt{\lambda_n}x) = \phi(x) - \frac{J_0(\omega x)}{J_0(\omega l)}$$

$$v_t(x,0) = \sum_{n=1}^{\infty} B_n \sqrt{\lambda_n} J_0(\sqrt{\lambda_n}x) = \psi(x).$$

For this eigenvalue problem, $m(x) = x$ and the inner product is given by

$$(f, g) = \int_0^l x f(x) g(x) \, dx.$$

Therefore, the coefficients are given by

$$A_n = \frac{\left(J_0(\sqrt{\lambda_n} x), \phi(x) - \frac{J_0(\omega x)}{J_0(\omega l)}\right)}{\left(J_0(\sqrt{\lambda_n} x), J_0(\sqrt{\lambda_n} x)\right)} = \frac{\int_0^l x J_0(\sqrt{\lambda_n} x)[\phi(x) - 1] \, dx}{\int_0^l x J_0^2(\sqrt{\lambda_n} x) \, dx}$$

and

$$B_n = \frac{1}{\sqrt{\lambda_n}} \frac{\left(J_0(\sqrt{\lambda_n} x), \psi\right)}{\left(J_0(\sqrt{\lambda_n} x), J_0(\sqrt{\lambda_n} x)\right)} = \frac{1}{\sqrt{\lambda_n}} \frac{\int_0^l x J_0(\sqrt{\lambda_n} x) \psi(x) \, dx}{\int_0^l x J_0^2(\sqrt{\lambda_n} x) \, dx}.$$

Therefore,

$$u(x, t) = \sum_{n=1}^{\infty} J_0(\sqrt{\lambda_n} x)[A_n \cos(\sqrt{\lambda_n} t) + B_n \sin(\sqrt{\lambda_n} t)] + \frac{J_0(\omega x)}{J_0(\omega l)} \cos(\omega t).$$

11.4.7 If λ is an eigenvalue with eigenfunction v, then $\overline{\lambda}$ is an eigenvalue with eigenfunction $\overline{v}$. We will show that $\lambda = \overline{\lambda}$, and, therefore, λ is real. We have $(pv^{(m)})^{(m)} = \lambda v$. Integrating by parts, we see that

$$\lambda \int_a^b v \overline{v} \, dx = \int_a^b (pv^{(m)})^{(m)} \overline{v} \, dx$$

$$= -\int_a^b (pv^{(m)})^{(m-1)} \overline{v}' \, dx + (pv^{(m)})^{(m-1)} \overline{v} \Big|_a^b.$$

By assumption, $v = v' = \ldots = v^{(m)} = 0$. Therefore, the boundary term vanishes. Proceeding in the same way, we see that

$$\lambda \int_a^b v \overline{v} \, dx = (-1)^m \int_a^b pv^{(m)} \overline{v}^{(m)} \, dx$$

$$= (-1)^m \int_a^b v^{(m)} (p\overline{v}^{(m)}) \, dx$$

$$= (-1)^m (-1)^m \int_a^b v (p\overline{v}^{(m)})^{(m)} \, dx + 0,$$

since all boundary terms vanish. Consequently,

$$\lambda \int_a^b v \overline{v} \, dx = \int_a^b v (p\overline{v}^{(m)})^{(m)} \, dx$$

$$= \int_a^b v \overline{\lambda} \overline{v} \, dx$$

$$= \overline{\lambda} \int_a^b v \overline{v} \, dx.$$

Therefore,

$$(\lambda - \overline{\lambda}) \int_a^b v \overline{v} \, dx = 0,$$

which can be rewritten as

$$(\lambda - \overline{\lambda}) \int_a^b |v|^2 \, dx = 0.$$

Since $\int_a^b |v|^2 \, ds \neq 0$, we must have $(\lambda - \overline{\lambda}) = 0$. Therefore, $\lambda = \overline{\lambda}$ which implies that λ is real.

Section 11.5

11.5.2. Suppose u is a C^2 solution of $u_{tt} = c^2 \Delta u$ on some bounded domain D, with Dirichlet or Neumann or Robin boundary condition. Then $u(\mathbf{x}, t)$, $u_{tt}(\mathbf{x}, t)$ and $\Delta u(\mathbf{x}, t)$ are each L^2 functions of $\mathbf{x}$ on D for all fixed t. By completeness of the eigenfunctions $v_n(\mathbf{x})$ of $-\Delta$ on D, we therefore have the expansions

$$u(\mathbf{x}, t) = \sum_{n=1}^{\infty} c_n(t) v_n(\mathbf{x})$$

$$u_{tt}(\mathbf{x}, t) = \sum_{n=1}^{\infty} d_n(t) v_n(\mathbf{x})$$

$$\Delta u(\mathbf{x}, t) = \sum_{n=1}^{\infty} e_n(t) v_n(\mathbf{x})$$

for each fixed t, where each series converges in the L^2 sense. The coefficients are given by

$$c_n(t) = \frac{1}{(v_n, v_n)} \iiint_D u(\mathbf{x}, t) v_n(\mathbf{x}) \, d\mathbf{x},$$

$$d_n(t) = \frac{1}{(v_n, v_n)} \iiint_D u_{tt}(\mathbf{x}, t) v_n(\mathbf{x}) \, d\mathbf{x} = c_n''(t),$$

and

$$e_n(t) = \frac{1}{(v_n, v_n)} \iiint_D \Delta u(\mathbf{x}, t) v_n(\mathbf{x}) \, d\mathbf{x}$$

$$= \frac{1}{(v_n, v_n)} \iiint_D u(\mathbf{x}, t) \Delta v_n(\mathbf{x}) \, d\mathbf{x}$$

$$= \frac{-\lambda_n}{(v_n, v_n)} \iiint_D u(\mathbf{x}, t) v_n(\mathbf{x}) \, d\mathbf{x}$$

$$= -\lambda_n c_n(t).$$

Since u is a solution of the wave equation we have

$$0 = u_{tt} - c^2 \Delta u = \sum_{n=1}^{\infty} (d_n(t) - c^2 e_n(t)) v_n(\mathbf{x}),$$

which implies, by completeness, that $d_n(t) = c^2 e_n(t)$. By the calculations above, this may be written $c_n''(t) = -c^2 \lambda_n c_n(t)$. As this holds for all t, and each λ_n is nonnegative, solving this ODE for c_n gives

$$c_n(t) = A_n \cos(c\sqrt{\lambda_n}\,t) + B_n \sin(c\sqrt{\lambda_n}\,t).$$

Section 11.6

11.6.3.

(a) To find upper bounds, we must consider regions contained within the ellipse. Fix $0 < x < 1$ and consider the rectangle $[-x, x] \times [-2\sqrt{1-x^2}, 2\sqrt{1-x^2}]$ inscribed in the ellipse. This rectangle has width $a = 2x$ and height $b = 4\sqrt{1-x^2}$, so its Dirichlet eigenvalues are

$$\frac{m^2\pi^2}{4x^2} + \frac{n^2\pi^2}{16(1-x^2)}.$$

The first eigenvalue clearly occurs when $m = n = 1$, so

$$\lambda_1(x) = \frac{\pi^2}{16}\left(\frac{4}{x^2} + \frac{1}{1-x^2}\right).$$

To find the best upper bound of this type, let $t = x^2$ and minimize over $t \in (0, 1)$.

$$\lambda_1(t) = \frac{\pi^2}{16}\left(\frac{4}{t} + \frac{1}{1-t}\right) \implies \lambda_1'(t) = \frac{\pi^2}{16}\left(-\frac{4}{t^2} + \frac{1}{(1-t)^2}\right).$$

Solving the quadratic equation, $t = 2/3$ is the critical point. Thus the best upper bound for λ_1 using rectangles with sides parallel to the coordinate axes is

$$\lambda_1 \leq \frac{9\pi^2}{16}.$$

Now the second eigenvalue is either

$$\lambda_2(x) = \frac{\pi^2}{16}\left(\frac{16}{x^2} + \frac{1}{1-x^2}\right) \quad \text{or} \quad \lambda_2(x) = \frac{\pi^2}{16}\left(\frac{4}{x^2} + \frac{4}{1-x^2}\right),$$

depending on which is smaller. The first expression is minimized at $x^2 = 4/5$, with a value of $25\pi^2/16$. The second expression is minimized at $x^2 = 1/2$, with a value of π^2. Thus the best upper bound of the second eigenvalue we can obtain in this way is

$$\lambda_2 \leq \pi^2.$$

(b) To obtain lower bounds, consider the rectangle $[-1, 1] \times [-2, 2]$. Its eigenvalues are

$$\frac{m^2\pi^2}{4} + \frac{n^2\pi^2}{16},$$

and thus we have the lower bounds

$$\lambda_1 \geq \frac{5\pi^2}{16} \quad \text{and} \quad \lambda_2 \geq \frac{\pi^2}{2}.$$

11.6.5 We proceed as in Example 11.6.2. $N(\lambda)$ is at most the volume of the ellipsoid

$$\frac{l^2}{a^2} + \frac{m^2}{b^2} + \frac{k^2}{c^2} \leq \frac{\lambda}{\pi^2}$$

in the first octant. Therefore,

$$N(\lambda) \leq \frac{1}{8} \cdot \frac{4\pi}{3} \frac{abc\lambda^{3/2}}{\pi^3} = \frac{abc\lambda^{3/2}}{6\pi^2}.$$

$N(\lambda)$ and this volume may differ by approximately the surface area of the ellipsoid, which is of order λ. Therefore,

$$\frac{abc\lambda^{3/2}}{6\pi^2} - C\lambda \leq N(\lambda) \leq \frac{abc\lambda^{3/2}}{6\pi^2}.$$

Substituting $\lambda = \lambda_n$ and $N(\lambda) = n$, we have

$$\frac{abc\lambda_n^{3/2}}{6\pi^2} - C\lambda_n \leq n \leq \frac{abc\lambda_n^{3/2}}{6\pi^2}.$$

Now dividing by n, we have

$$\frac{abc\lambda_n^{3/2}}{6\pi^2 n} - C\frac{\lambda_n}{n} \leq 1 \leq \frac{abc\lambda_n^{3/2}}{6\pi^2 n}.$$

Therefore, we conclude that

$$\lim_{n \to \infty} \frac{\lambda_n^{3/2}}{n} = \frac{6\pi^2}{abc}.$$

11.6.7.

(a) Suppose $v_1(\mathbf{x}_0) = 0$ for some $\mathbf{x}_0$ in D. Now suppose $v_1(\mathbf{x}) \leq 0$ for all $\mathbf{x}$ in D. Then $\Delta v_1 = -\lambda_1 v_1 \geq 0$ in D, and therefore v_1 is subharmonic and therefore cannot attain a maximum within D. However, since $v_1 \leq 0$ on D and $v_1(\mathbf{x}_0) = 0$, v_1 attains its maximum at $\mathbf{x}_0 \in D$. This contradiction implies that v_1 is positive somewhere in D. Next suppose $v_1(\mathbf{x}) \geq 0$ for all $\mathbf{x}$ in D. Then $\Delta(-v_1) = \lambda_1 v_1 \geq 0$ in D, so $-v_1$ is subharmonic and therefore cannot attain a maximum on D. But since $-v_1 \leq 0$ on D and $-v_1(\mathbf{x}_0) = 0$, $-v_1$ attains its maximum at $\mathbf{x}_0 \in D$. This contradiction implies that v_1 is negative somewhere in D.

(b) It is clear that $\||v_1|\| = \|v_1\|$. Next, $|v_1| = v^+ - v^-$, so that

$$\nabla|v_1| = \nabla v^+ - \nabla v^- = \begin{cases} \nabla v_1 & \text{on } D^+ \\ -\nabla v_1 & \text{on } D^- \end{cases}.$$

So $|\nabla|v_1|| = |\nabla v_1|$ on D^+ and D^-. Thus

$$\|\nabla|v_1|\|^2 = \iiint_D |\nabla|v_1||^2 \, d\mathbf{x}$$

$$= \iiint_{D^+} |\nabla|v_1||^2 \, d\mathbf{x} + \iiint_{D^-} |\nabla|v_1||^2 \, d\mathbf{x}$$

$$= \iiint_{D^+} |\nabla v_1|^2 \, d\mathbf{x} + \iiint_{D^-} |\nabla v_1|^2 \, d\mathbf{x}$$

$$= \iiint_D |\nabla|v_1||^2 \, d\mathbf{x}$$

$$= \|\nabla v_1\|^2.$$

Thus the Rayleigh quotients of v_1 and $|v_1|$ are the same, so $|v_1|$ is also an eigenfunction with eigenvalue λ_1.

(c) Since $|v_1|$ is an eigenfunction, $\Delta(-|v_1|) = \lambda_1|v_1| \geq 0$, so $-|v_1|$ is subharmonic, and therefore cannot attain a maximum on D, by Exercise 7.4.25. But $-|v_1| \leq 0$ on D and $-|v_1(\mathbf{x}_0)| = 0$, so $-|v_1|$ attains its maximum at $\mathbf{x}_0 \in D$. This contradiction shows that the initial assumption in part (a) that $v_1(\mathbf{x}_0) = 0$ was false. Hence v_1 never vanishes, and by continuity we have either $v_1 > 0$ on D or $v_1 < 0$ on D.

(d) Suppose u is an eigenfunction with eigenvalue λ_1, and let

$$w = u - \frac{(u, v_1)}{(v_1, v_1)} v_1.$$

Then $(w, v_1) = 0$. Since $-\Delta w = \lambda_1 w$, either $w \equiv 0$, or w is an eigenfunction with eigenvalue λ_1. If the latter were the case, then part (c) implies that $w > 0$ on D or $w < 0$ on D. Since the same is true of v_1, the product wv_1 is either positive everywhere on D or negative everywhere on D. Hence

$$(w, v_1) = \iiint_D w(\mathbf{x})v_1(\mathbf{x}) \, d\mathbf{x}$$

cannot possibly vanish. This contradiction implies that $w \equiv 0$ on D. Hence

$$u = \frac{(u, v_1)}{(v_1, v_1)} v_1$$

is a scalar multiple of v_1. Thus the eigenspace of λ_1 has dimension 1, so λ_1 is a simple eigenvalue.

Chapter 12

Section 12.1

12.1.1. Let F be the distribution associated with f. That is, let F be defined by

$$(F, \phi) = \int_{-\infty}^{\infty} f(x)\phi(x)\, dx.$$

First, we check linearity.

$$
\begin{aligned}
(F, a\phi + b\psi) &= \int_{-\infty}^{\infty} f(x)[a\phi(x) + b\psi(x)]\, dx \\
&= a \int_{-\infty}^{\infty} f(x)\phi(x)\, dx + b \int_{-\infty}^{\infty} f(x)\phi(x)\, dx \\
&= a(F, \phi) + b(F, \psi).
\end{aligned}
$$

Second, we check continuity. Let $\{\phi_n\}$ be a sequence of functions which vanishes outside a common interval I and converges uniformly to ϕ. Fix $\epsilon > 0$. We see that

$$
\begin{aligned}
|(F_f, \phi_n) - (F_f, \phi)| &= \left| \int_I f(x)(\phi_n - \phi)\, dx \right| \\
&\leq \max_{x \in I} |\phi_n - \phi| \int_I |f(x)|\, dx.
\end{aligned}
$$

By assumption, the integral is finite. Therefore, since ϕ_n converges uniformly to ϕ, we have

$$\max_{x \in I} |\phi_n - \phi| \int_I |f(x)|\, dx < \epsilon$$

for $n \geq N$ sufficiently large.

12.1.5. Let $u(x, t) = H(x - ct)$, and let $\phi(x, t)$ be any test function. Then

$$(u, \phi) = \int_{-\infty}^{\infty} \int_{ct}^{\infty} \phi(y, t)\, dy\, dt$$

and

$$
\begin{aligned}
(u_{tt}, \phi) &= -(u_t, \phi_t) = +(u, \phi_{tt}) \\
&= \int_{-\infty}^{\infty} \int_{ct}^{\infty} \phi_{tt}(y, t)\, dy\, dt = \int_{-\infty}^{\infty} \int_{-\infty}^{y/c} \phi_{tt}(y, t)\, dt\, dy \\
&= \int_{-\infty}^{\infty} \phi_t(y, y/c)\, dy = c \int_{-\infty}^{\infty} \phi_t(cs, s)\, ds
\end{aligned}
$$

and

$$(u_{xx}, \phi) = -(u_x, \phi_x) = +(u, \phi_{xx})$$
$$= \int_{-\infty}^{\infty} \int_{ct}^{\infty} \phi_{xx}(y,t)\, dy\, dt = -\int_{-\infty}^{\infty} \phi_x(ct,t)\, dt.$$

Therefore

$$(u_{tt} - c^2 u_{xx}, \phi) = \int_{-\infty}^{\infty} c\phi_t(cs,s) + c^2 \phi_x(cs,s)\, ds$$
$$= c\int_{-\infty}^{\infty} \frac{d}{ds}[\phi(cs,s)]\, ds$$
$$= c\phi(cs,s)\Big|_{-\infty}^{\infty} = 0.$$

Since this holds for all test functions ϕ, this means that $u_{tt} - c^2 u_{xx} = 0$ in the sense of distributions.

12.1.6. Equation (12.1.17) holds pointwise, so multiplying by $\phi'(x)$ gives

$$\sum_{n \text{ odd}} \frac{4}{n\pi} \phi'(x) \sin nx = \begin{cases} \phi'(x) & \text{for } 0 < x < \pi \\ -\phi'(x) & \text{for } -\pi < x < 0 \end{cases}.$$

By Exercise 5.4.11, we may integrate term by term and use the fact that ϕ vanishes near $\pm\pi$ to get

$$\sum_{n \text{ odd}} \int_{-\pi}^{\pi} \frac{4}{n\pi} \phi'(x) \sin nx\, dx = \int_0^{\pi} \phi'(x)\, dx + \int_{-\pi}^0 -\phi'(x)\, dx = -2\phi(0).$$

On the other hand, integration by parts gives

$$\int_{-\pi}^{\pi} \frac{4}{n\pi} \phi'(x) \sin nx\, dx = \frac{4}{n\pi} \phi(x) \sin nx \Big|_{-\pi}^{\pi} - \frac{4}{\pi} \int_{-\pi}^{\pi} \phi(x) \cos nx\, dx = -\frac{4}{\pi} \int_{-\pi}^{\pi} \phi(x) \cos nx\, dx.$$

Combining the last two equations and multiplying by $-\pi/4$, we deduce

$$\sum_{n \text{ odd}} \int_{-\pi}^{\pi} \phi(x) \cos nx\, dx = \frac{\pi\phi(0)}{2}.$$

12.1.7. Let F_n be the distribution associated with f_n such that

$$(F_n, \phi) = \int_{-\infty}^{\infty} f_n(x)\phi(x)\, dx.$$

Let F be the distribution associated with f such that

$$(F, \phi) = \int_{-\infty}^{\infty} f(x)\phi(x)\, dx.$$

176

To show $F_n \to F$ weakly, we need to show that

$$(F_n, \phi) \to (F, \phi)$$

for all $\phi \in \mathscr{D}$. That is, we need to show that for $\epsilon > 0$, there exists an $N > 0$ such that

$$|(F_n, \phi) - (F, \phi)| < \epsilon$$

for $n \geq N$ sufficiently large. Now since $\phi \in \mathscr{D}$, ϕ has bounded L^2 norm, so using the Schwarz inequality (see Exercise 5.5.2), we have

$$|(F_n, \phi) - (F, \phi)| = \left| \int_{-\infty}^{\infty} [f_n(x) - f(x)] \phi(x) \, dx \right|$$
$$\leq \|f_n - f\| \|\phi\|.$$

Since $f_n \to f$ in L^2, it follows that there is some $N > 0$ such that

$$\|f_n - f\| \|\phi\| < \epsilon$$

for all $n \geq N$.

12.1.10. To show that the mapping

$$\phi \to \iint_S \phi \, dS$$

is a distribution, we must verify that it is linear and continuous. As for the linearity,

$$\iint_S (a\phi + b\psi) \, dS = a \iint_S \phi \, dS + b \iint_S \psi \, dS.$$

As for the continuity, let $\{\psi_n\}$ be a sequence of test functions vanishing outside some fixed ball $D = \{|\mathbf{x}| < a\}$, and let ϕ_n converge uniformly to ϕ, together with all of their derivatives. Let $\epsilon > 0$. Then the uniform convergence implies that there is some N such that $|\phi_n(\mathbf{x}) - \phi(\mathbf{x})| < \epsilon / 4\pi a^2$ for all $\mathbf{x}$ and all $n \geq N$. It then follows that

$$\left| \iint_S \phi_n \, dS - \iint_S \phi \, dS \right| \leq \iint_{S \cap D} |\phi_n - \phi| \, dS$$
$$\leq 4\pi a^2 \max_{\mathbf{x} \in D} |\phi_n(\mathbf{x}) - \phi(\mathbf{x})|$$
$$< \epsilon$$

for all $n \geq N$. Therefore $\iint_S \phi_n \, dS \to \iint_S \phi \, dS$.

12.1.12. Given a test function ϕ, we need to show that

$$\lim_{a \to 0} (\chi_a, \phi) = (\delta, \phi).$$

Since $\int_{-a}^{a} \chi_a(a)\,dx = 1$ for all a, we have

$$(\chi_a, \phi) - (\delta, \phi) = \int_{-\infty}^{\infty} \chi_a(x)\phi(x)\,dx - \phi(0)$$

$$= \int_{-a}^{a} \chi_a(x)(\phi(x) - \phi(0))\,dx$$

$$= \frac{1}{2a} \int_{-a}^{a} (\phi(x) - \phi(0))\,dx.$$

Now given any $\epsilon > 0$, by the continuity of ϕ, there exists an $a > 0$ such that $|x| < a$ implies $|\phi(x) - \phi(0)| < \epsilon$. For such a we then have

$$|(\chi_a, \phi) - (\delta, \phi)| \leq \frac{1}{2a} \int_{-a}^{a} |\phi(x) - \phi(0)|\,dx < \frac{1}{2a} \int_{-a}^{a} \epsilon\,dx = \epsilon,$$

which establishes the desired limit.

Section 12.2

12.2.2. The initial data is exactly the initial data for the Riemann function, multiplied by V and shifted by x_0. So

$$u(x,t) = VS(x - x_0, t) = \begin{cases} \frac{V}{2c} & |x - x_0| < ct \\ 0 & |x - x_0| > ct. \end{cases}$$

12.2.4. Since

$$S(x,t) = \begin{cases} \frac{1}{2c} & |x| < ct \\ 0 & |x| > ct \end{cases}$$

for each fixed t, we can regard S as the distribution in x given by

$$(S, \phi) = \frac{1}{2c} \int_{-ct}^{ct} \phi(x)\,dx.$$

Then, by the Fundamental Theorem of Calculus,

$$\left(\frac{\partial S}{\partial t}, \phi\right) = \frac{1}{2c}\left[c\phi(ct) + c\phi(-ct)\right] = \frac{1}{2}\left[\phi(ct) + \phi(-ct)\right],$$

which implies

$$\frac{\partial S}{\partial t}(x,t) = \frac{1}{2}[\delta(x - ct) + \delta(x + ct)].$$

Next let $R = \partial S/\partial t$. Then R satisfies $R_{tt} = c^2 R_{xx}$, with initial data $R(x,0) = S_t(x,0) = \delta(x)$ and $R_t(x,0) = S_{tt}(x,0) = c^2 \Delta S(x,0) = 0$. Thus R solves the wave equation with $R(x,0) = \delta(x)$ and $R_t(x,0) = 0$.

12.2.6. By Duhamel's Principle (9.3.11), the solution of the inhomogeneous wave equation with homogeneous initial data

$$u_{tt} = c^2 \Delta u + f$$
$$u(\mathbf{x}, 0) = 0$$
$$u_t(\mathbf{x}, 0) = 0$$

is

$$u(\mathbf{x}, t) = \int_0^t \int_{\mathbf{R}^n} S(\mathbf{x} - \mathbf{y}, t - s) f(\mathbf{y}, s) d\mathbf{y} \, ds.$$

As shown in Section 12.2, the solution of

$$u_{tt} = c^2 \Delta u$$
$$u(\mathbf{x}, 0) = 0$$
$$u_t(\mathbf{x}, 0) = \psi(\mathbf{x})$$

is

$$u(\mathbf{x}, t) = \int_{\mathbf{R}^n} S(\mathbf{x} - \mathbf{y}, t) \psi(\mathbf{y}) d\mathbf{y}.$$

As in Exercise 12.2.4, the solution of

$$u_{tt} = c^2 \Delta u$$
$$u(\mathbf{x}, 0) = \phi(\mathbf{x})$$
$$u_t(\mathbf{x}, 0) = 0$$

is

$$u(\mathbf{x}, t) = \frac{\partial}{\partial t} \int_{\mathbf{R}^n} S(\mathbf{x} - \mathbf{y}, t) \phi(\mathbf{y}) d\mathbf{y}.$$

Adding these, the solution of

$$u_{tt} = c^2 \Delta u + f$$
$$u(\mathbf{x}, 0) = \phi(\mathbf{x})$$
$$u_t(\mathbf{x}, 0) = \psi(\mathbf{x})$$

is

$$u(\mathbf{x}, t) = \int_{\mathbf{R}^n} S(\mathbf{x} - \mathbf{y}, t) \psi(\mathbf{y}) d\mathbf{y} + \frac{\partial}{\partial t} \int_{\mathbf{R}^n} S(\mathbf{x} - \mathbf{y}, t) \phi(\mathbf{y}) d\mathbf{y}$$
$$+ \int_0^t \int_{\mathbf{R}^n} S(\mathbf{x} - \mathbf{y}, t - s) f(\mathbf{y}, s) d\mathbf{y} \, ds.$$

12.2.8.

(a) This is the composition of the delta function with the function $w = \psi(r) = a^2 - r^2$. The inverse of this function is $r = \psi^{-1}(w) = \sqrt{a^2 - w}$ for $r > 0$, and $\psi'(r) = -2r = -2\sqrt{a^2 - w}$. For any test function $\phi(r)$ we have, by the definition in the text,

$$(\delta(a^2 - r^2), \phi) = \left(\delta(w), \frac{\phi(\sqrt{a^2 - w})}{2\sqrt{a^2 - w}} \right) = \frac{\phi(a)}{2a}$$

for $a > 0$. On the other hand, with $z = a - r$,

$$(\delta(a - r), \phi) = \left(\delta(z), \frac{\phi(a - z)}{1}\right) = \phi(a),$$

so that $(\delta(a^2 - r^2), \phi) = \left(\frac{1}{2a}\delta(a - r), \phi\right)$ for all ϕ.

(b) By the result of part (a), and the formula in the text,

$$S(\mathbf{x}, t) = \frac{1}{4\pi c^2 t}\delta(ct - |\mathbf{x}|) = \frac{1}{4\pi c^2 t} 2ct\delta((ct)^2 - |\mathbf{x}|^2) = \frac{1}{2\pi c}\delta(c^2 t^2 - |\mathbf{x}|^2).$$

12.2.10. By Duhamel's principle (9.3.11) and (12.2.12), we have

$$u(x_0, y_0, t) = \frac{1}{2\pi c} \int_0^t \iint_D \frac{\delta(x)\delta(y)f(s)}{(c^2(t - s)^2 - (x - x_0)^2 - (y - y_0)^2)^{1/2}} \, dy \, dx \, ds$$

where D is the ball $D = \{(x - x_0)^2 + (y - y_0)^2 \le c^2(t - s)^2$. Let $r = \sqrt{x_0^2 + y_0^2}$. Notice that the ball D intersects the origin whenever $c(t - s) \ge r$. Thus, if $r > ct$, the integrand vanishes, and $u(x_0, y_0, t) = 0$. Otherwise, when $r \le ct$ we have

$$u(x_0, y_0, t) = \frac{1}{2\pi c} \int_0^{t - r/c} \frac{f(s)}{(c^2(t - s)^2 - r^2)^{1/2}} \, ds.$$

12.2.12. As in Section 10.1 (see Example 10.1.1), we separate variables to find solutions of the form

$$\sin \frac{m\pi x}{a} \sin \frac{n\pi y}{b} \exp\left[-\left(\frac{m^2}{a^2} + \frac{n^2}{b^2}\right)\pi^2 kt\right].$$

Write u as a series

$$u(x, y, t) = \sum_{m=1}^{\infty} \sum_{n=1}^{\infty} A_{mn} \sin \frac{m\pi x}{a} \sin \frac{n\pi y}{b} \exp\left[-\left(\frac{m^2}{a^2} + \frac{n^2}{b^2}\right)\pi^2 kt\right].$$

The initial condition requires

$$M\delta(x - a/2)\delta(y - b/2) = \sum_{m=1}^{\infty} \sum_{n=1}^{\infty} \sin \frac{m\pi x}{a} \sin \frac{n\pi y}{b},$$

so that

$$\begin{aligned}
A_{mn} &= \frac{4}{ab} \int_0^b \int_0^a M\delta(x - a/2)\delta(y - b/2) \sin \frac{m\pi x}{a} \sin \frac{n\pi y}{b} \, dx \, dy \\
&= \frac{4M}{ab} \sin \frac{m\pi}{2} \sin \frac{n\pi}{2} \quad \text{(by the definition of the } \delta \text{ function)} \\
&= \frac{4M}{ab}\left(\frac{1}{2}\cos((m - n)\pi/2) - \frac{1}{2}\cos((m + n)\pi/2)\right) \\
&= \left\{ \begin{array}{cl}
0 & \text{if either } m \text{ or } n \text{ is even} \\
\dfrac{2M\left((-1)^{\frac{m-n}{2}} - (-1)^{\frac{m+n}{2}}\right)}{ab} & \text{if both } m \text{ and } n \text{ are odd}
\end{array} \right\}.
\end{aligned}$$

Section 12.3

12.3.1.

(5) For the delta function we must interpret the formula (12.3.4) that defines the Fourier transform as the distribution $\delta(x)$ acting on e^{-ikx}. That is,

$$\int_{-\infty}^{\infty} \delta(x)e^{-ikx}\,dx = \left(\delta, e^{-ikx}\right) = e^0 = 1.$$

(6) For the square pulse,

$$\int_{-\infty}^{\infty} H(a-|x|)e^{-ikx}\,dx = \int_{-a}^{a} e^{-ikx}\,dx = \left.\frac{e^{-ikx}}{-ik}\right|_{-a}^{a} = \frac{e^{-ika}-e^{ika}}{-ik} = \frac{2\sin ak}{k}.$$

(7) For the exponential,

$$\int_{-\infty}^{\infty} e^{-a|x|}e^{-ikx}\,dx = \int_{-\infty}^{0} e^{(a-ik)x}\,dx + \int_{0}^{\infty} e^{(-a-ik)x}\,dx$$

$$= \lim_{b\to-\infty} \left.\frac{e^{(a-ik)x}}{a-ik}\right|_{b}^{0} + \lim_{b\to\infty} \left.\frac{e^{(-a-ik)x}}{-a-ik}\right|_{0}^{b}$$

$$= \frac{1}{a-ik} - \frac{1}{-a-ik} = \frac{2a}{a^2+k^2}$$

for $a > 0$.

(10) For the constant function 1, first notice that equations (12.3.3) and (12.3.4) imply that taking the Fourier transform of a function $f(x)$ twice gives $2\pi f(-x)$. Thus since 1 is the Fourier transform of δ, the Fourier transform of 1 must be $2\pi\delta(-k) = 2\pi\delta(k)$.

[Strictly speaking, a proper justification of this fact requires the development of the Fourier transform of distributions (as in [Fo], for instance).]

12.3.2.

(i) Integrate by parts and assume that $f(x) \to 0$ as $|x| \to \infty$:

$$\int_{-\infty}^{\infty} \frac{df}{dx} e^{-ikx}\,dx = \left. f(x)e^{-ikx}\right|_{-\infty}^{\infty} - \int_{-\infty}^{\infty} f(x)(-ik)e^{-ikx}\,dx$$

$$= ik \int_{-\infty}^{\infty} f(x)e^{-ikx}\,dx = ikF(k).$$

(iii) Make the change of variable $y = x - a$:

$$\int_{-\infty}^{\infty} f(x-a)e^{-ikx}\,dx = \int_{-\infty}^{\infty} f(y)e^{-ik(y+a)}\,dy = e^{-ika}\int_{-\infty}^{\infty} f(y)e^{-iky}\,dy = e^{-ika}F(k).$$

181

(vi) Make the change of variable $y = ax$. For $a > 0$,

$$\int_{-\infty}^{\infty} f(ax)e^{-ikx}\, dx = \frac{1}{a}\int_{-\infty}^{\infty} f(y)e^{-i(k/a)y}\, dy = \frac{1}{a}F\left(\frac{k}{a}\right),$$

while for $a < 0$,

$$\int_{-\infty}^{\infty} f(ax)e^{-ikx}\, dx = \frac{1}{a}\int_{+\infty}^{-\infty} f(y)e^{-i(k/a)y}\, dy = -\frac{1}{a}\int_{-\infty}^{+\infty} f(y)e^{-i(k/a)y}\, dy$$

$$= -\frac{1}{a}F\left(\frac{k}{a}\right).$$

12.3.4.

(a) Making the change of variable $z = x - y$, $dz = -dy$ gives

$$(f * g)(x) = \int_{-\infty}^{\infty} f(x - y)g(y)\, dy$$

$$= -\int_{\infty}^{-\infty} f(z)g(x - z)\, dz$$

$$= \int_{-\infty}^{\infty} g(x - z)f(z)\, dz$$

$$= (g * f)(x).$$

12.3.7.

(a) Since f vanishes for large $|x|$, the summation is a finite sum for each fixed x and therefore converges. To see that g is periodic, compute

$$g(x + 2\pi) = \sum_{n=-\infty}^{\infty} f(x + 2\pi + 2\pi n) = \sum_{m=-\infty}^{\infty} f(x + 2\pi m) = g(x),$$

where we have substituted $m = n + 1$.

(b)

$$c_m = \frac{1}{2\pi}\int_{-\pi}^{\pi} g(x)e^{-imx}\, dx$$

$$= \sum_{n=-\infty}^{\infty} \frac{1}{2\pi}\int_{-\pi}^{\pi} f(x + 2\pi n)e^{-imx}\, dx$$

$$= \sum_{n=-\infty}^{\infty} \frac{1}{2\pi}\int_{-\pi+2\pi n}^{\pi+2\pi n} f(y)e^{-im(y-2\pi n)}\, dy$$

$$= \sum_{n=-\infty}^{\infty} \frac{1}{2\pi}\int_{\pi(2n-1)}^{\pi(2n+1)} f(y)e^{-imy}\, dy \qquad \text{(since } e^{2\pi ni} = 1\text{)}$$

$$= \frac{1}{2\pi}\int_{-\infty}^{\infty} f(y)e^{-imy}\, dy = \frac{F(m)}{2\pi}$$

182

(c) The Fourier series for g is

$$g(x) = \sum_{m=-\infty}^{\infty} c_m e^{-imx},$$

so at $x = 0$, $g(0) = \sum_{m=-\infty}^{\infty} c_m$, and by part (b) and the definition of g, this becomes

$$\sum_{n=-\infty}^{\infty} f(2\pi n) = \sum_{n=-\infty}^{\infty} \frac{F(n)}{2\pi}.$$

12.3.8. For $k \neq 0$,

$$
\begin{aligned}
\widehat{\chi_a}(k) &= \int_{-\infty}^{\infty} e^{-ixk} \chi_a(x)\, dx \\
&= \frac{1}{2a} \int_{-a}^{a} e^{-ixk}\, dx \\
&= \frac{1}{2a} \left. \frac{e^{-ixk}}{-ik} \right|_{x=-a}^{x=a} \\
&= \frac{e^{-iak} - e^{-iak}}{2ai} \\
&= \frac{\sin(ak)}{ak}.
\end{aligned}
$$

For $k = 0$, $\widehat{\chi_a}(k) = 1$.

Now let ϕ be a test function. Suppose ϕ vanishes outside $[-R, R]$, and $|\phi(k)| \leq M$ for all k. Since

$$\lim_{s \to 0} \frac{\sin(s)}{s} = 1,$$

it follows that for any $\epsilon > 0$ there is some $\delta > 0$ such that

$$\left| \frac{\sin(s)}{s} - 1 \right| < \frac{\epsilon}{2MR}$$

whenever $0 < |s| \leq \delta$. Thus, whenever $0 < |a| \leq \delta/R$ we have

$$\max_{0 < |k| \leq R} \left| \frac{\sin(ak)}{ak} - 1 \right| < \frac{\epsilon}{2MR}.$$

Therefore

$$\max_{|k| \leq R} |\widehat{\chi_a}(k) - 1| < \frac{\epsilon}{2MR}$$

so that

$$
\begin{aligned}
|(\widehat{\chi_a}(k), \phi) - (1, \phi)| &= \left| \int_{-R}^{R} [\widehat{\chi_a}(k) - 1]\, \phi(k)\, dk \right| \\
&\leq 2RM \max_{|k| \leq R} |\widehat{\chi_a}(k) - 1| < \epsilon.
\end{aligned}
$$

Therefore $\hat{\chi}_a \to 1$ weakly as $a \to 0$.

12.3.9. Taking the Fourier transform of both sides of the ODE gives

$$(k^2 + a^2)\hat{u}(k) = 1 \qquad \Longrightarrow \qquad \hat{u}(k) = \frac{1}{k^2 + a^2}$$

and by equation (12.3.7), this implies $u(x) = \frac{1}{2a}e^{-a|x|}$.

Section 12.4

12.4.1. Solution 1. Taking the Fourier transform of the PDE and initial condition with respect to x leads to

$$\hat{u}_t(k, t) = (-\kappa k^2 + i\mu k)\hat{u}(k, t) \qquad u(k, 0) = \hat{\phi}(k).$$

Solving the ODE gives $\hat{u}(k, t) = \hat{\phi}(k)e^{-\kappa k^2 t + i\mu k t}$. Now suppose w is a solution of the regular heat equation $w_t = \kappa w_{xx}$ with initial data ϕ. Then by the same reasoning we would have

$$\hat{w}(k, t) = \hat{\phi}(k)e^{-\kappa k^2 t} = e^{-i\mu t k}\hat{u}(k, t).$$

By formula (iii) in Section 12.3, it follows that $u(x, t) = w(x + \mu t, t)$. But w is given explicitly by the convolution of the heat kernel with ϕ (see (2.4.8)), so

$$u(x, t) = \frac{1}{\sqrt{4\pi\kappa t}}\int_{-\infty}^{\infty} e^{-(x+\mu t - y)^2/4\kappa t}\phi(y)\,dy.$$

Solution 2. Let S denote the solution of $S_t = \kappa S_{xx} + \mu S_x$ with initial data $S(x, 0) = \delta(x)$. Then $\hat{S}_t = (-\kappa k^2 + i\mu k)\hat{S}$ and $\hat{S}(k, 0) = 1$. The solution of this ODE is $\hat{S}(k, t) = e^{-\kappa k^2 t + i\mu k t}$. By property (iii) of the Fourier transform, it follows that $S(x, t) = G(x + \mu t, t)$, where $\hat{G}(k, t) = e^{-\kappa k^2 t}$. Using property (vi) together with formula (12.3.11) it follows that $G(x, t) = \frac{1}{\sqrt{4\pi\kappa t}}e^{-x^2/4\kappa t}$, and thus $S(x, t) = \frac{1}{\sqrt{4\pi\kappa t}}e^{-(x+\mu t)^2/4\kappa t}$. The solution with initial data ϕ is therefore

$$u(x, t) = S * \phi = \frac{1}{\sqrt{4\pi\kappa t}}\int_{-\infty}^{\infty} e^{-(x+\mu t - y)^2/4\kappa t}\phi(y)\,dy.$$

12.4.3. Motivated by the method used at the end of Section 6.1 to derive the fundamental solution of Laplace's equation, we first suppose u is a radial solution of the homogeneous equation $-\Delta u + m^2 u = 0$. Then $-u_{rr} - \frac{2}{r}u_r + m^2 u = 0$. Substituting $v = ru$ this becomes $v_{rr} = m^2 v$, so $v = Ce^{\pm mr}$. Since we want a bounded solution, we take the negative exponent, and therefore $u = \frac{Ce^{-mr}}{r}$. Notice that u satisfies the homogeneous PDE, except at the origin. Next let's compute the Fourier transform of $f(\mathbf{x}) = \frac{e^{-m|\mathbf{x}|}}{|\mathbf{x}|}$. Fix $\mathbf{k} \in \mathbf{R}^3$ and define spherical coordinates in $\mathbf{R}^3$ such that $\mathbf{k}$ is on the positive z-axis. If we denote $r = |\mathbf{x}|$ and $\rho = |\mathbf{k}|$, then $\mathbf{x} \cdot \mathbf{k} = |\mathbf{x}||\mathbf{k}|\cos\phi = r\rho\cos\phi$. Thus

$$\hat{f}(\mathbf{k}) = \iiint_{\mathbf{R}^3} \frac{e^{-m|\mathbf{x}|}}{|\mathbf{x}|}e^{-i\mathbf{x}\cdot\mathbf{k}}\,d\mathbf{x}$$

$$= \int_0^{2\pi}\int_0^{\pi}\int_0^{\infty} \frac{e^{-mr}}{r}e^{-ir\rho\cos\theta}r^2\sin\theta\,dr\,d\theta\,d\phi.$$

184

Computing the θ and ϕ integrals first gives

$$2\pi \int_0^\infty r e^{-mr} \left[\frac{e^{-ir\rho\cos\theta}}{ir\rho} \right]_0^\pi dr = \frac{2\pi}{i\rho} \int_0^\infty e^{r(-m+i\rho)} - e^{r(-m-i\rho)} \, dr$$

$$= \frac{2\pi}{i\rho} \left[\frac{e^{r(-m+i\rho)}}{-m+i\rho} - \frac{e^{r(-m-i\rho)}}{-m-i\rho} \right]_0^\infty$$

$$= \frac{-2\pi}{i\rho} \left[\frac{1}{i\rho - m} + \frac{1}{i\rho + m} \right]$$

$$= \frac{4\pi}{(\rho^2 + m^2)}$$

$$= \frac{4\pi}{|\mathbf{k}|^2 + m^2}.$$

Finally, taking the Fourier transform of the original PDE $-\Delta u + m^2 u = \delta$ gives

$$(|\mathbf{k}|^2 + m^2)\hat{u} = 1 \qquad \Longrightarrow \qquad \hat{u}(\mathbf{k}) = \frac{1}{|\mathbf{k}|^2 + m^2}.$$

So

$$u(\mathbf{x}) = \frac{1}{4\pi} f(\mathbf{x}) = \frac{e^{-m|\mathbf{x}|}}{4\pi |\mathbf{x}|}.$$

12.4.5.

(a) Taking the Fourier transform in the x and y variables, we have

$$(ik)^2 \widehat{u} + (il)^2 \widehat{u} + \widehat{u}_{zz} = 0.$$

(Note that we cannot take the Fourier transform in the z variable, because z is restricted to be positive.) The solutions of this ODE are given by

$$\widehat{u}(k, l, z) = C_1 e^{\sqrt{k^2 + l^2}\, z} + C_2 e^{-\sqrt{k^2 + l^2}\, z}.$$

As we don't want the solution to go to ∞ as $|k|, |l| \to +\infty$, we take

$$\widehat{u}(k, l, z) = C e^{-\sqrt{k^2 + l^2}\, z}.$$

Now the boundary condition $u(x, y, 0) = \delta(x, y)$ implies $\widehat{u}(k, l, 0) = 1$. Therefore $C = 1$ and

$$u(x, y, z) = \frac{1}{(2\pi)^2} \iint_{\mathbf{R}^2} e^{ixk + iyl} \widehat{u}(k, l, z) \, dk \, dl$$

$$= \frac{1}{(2\pi)^2} \iint_{\mathbf{R}^2} e^{-ixk + iyl} e^{-\sqrt{k^2 + l^2}\, z} \, dk \, dl,$$

as claimed.

185

(b) Now we make a change of variables, rewriting this integral in polar coordinates in the (k, l)-plane. We define our transformation such that

$$k = \rho \cos \theta \qquad l = \rho \sin \theta$$

where θ is the angle between (k, l) and (x, y) (which is fixed) and $\rho = \sqrt{k^2 + l^2}$. In addition, let $r = \sqrt{x^2 + y^2}$. With this change of variables, our integral can be rewritten as

$$u(x, y, z) = \frac{1}{4\pi^2} \int_0^{2\pi} \int_0^\infty e^{ir\rho \cos \theta} e^{-\rho z} \rho \, d\rho \, d\theta.$$

(c) Integrating by parts, we have

$$u(x, y, z) = \frac{1}{4\pi^2} \int_0^{2\pi} \int_0^\infty e^{\rho(ir \cos \theta - z)} \rho \, d\rho \, d\theta$$

$$= \frac{1}{4\pi^2} \int_0^{2\pi} \frac{e^{\rho(ir \cos \theta - z)} \rho}{ir \cos \theta - z} \bigg|_{\rho=0}^{\rho=\infty} d\theta - \frac{1}{4\pi^2} \int_0^{2\pi} \int_0^\infty \frac{e^{\rho(ir \cos \theta - z)}}{ir \cos \theta - z} \, d\rho \, d\theta.$$

Because $z > 0$, the first remaining term vanishes. Therefore, we have

$$u(x, y, z) = -\frac{1}{4\pi^2} \int_0^{2\pi} \int_0^\infty \frac{e^{\rho(ir \cos \theta - z)}}{ir \cos \theta - z} \, d\rho \, d\theta$$

$$= -\frac{1}{4\pi^2} \int_0^{2\pi} \frac{e^{\rho(ir \cos \theta - z)}}{(ir \cos \theta - z)^2} \bigg|_{\rho=0}^{\rho=\infty} d\theta$$

$$= \frac{1}{4\pi^2} \int_0^{2\pi} \frac{1}{(ir \cos \theta - z)^2} \, d\theta.$$

Then, using an integral table, we conclude that

$$u(x, y, z) = \frac{2\pi z}{(r^2 + z^2)^{3/2}}.$$

Section 12.5

12.5.1.

(2)

$$F(s) = \int_0^\infty e^{at} e^{-st} \, dt = \left[\frac{e^{at-st}}{a-s} \right]_{t=0}^{t=\infty} = \frac{1}{s-a} \quad \text{for} \quad s > a$$

(3) Integrate by parts twice.

$$F(s) = \int_0^\infty (\cos \omega t) e^{-st} \, dt$$

$$= \left[-\frac{1}{s} (\cos \omega t) e^{-st} \right]_{t=0}^{t=\infty} - \frac{\omega}{s} \int_0^\infty (\sin \omega t) e^{-st} \, dt$$

$$= \frac{1}{s} + \left[\frac{\omega}{s^2} (\sin \omega t) e^{-st} \right]_{t=0}^{t=\infty} - \frac{\omega^2}{s^2} \int_0^\infty (\cos \omega t) e^{-st} \, dt$$

$$= \frac{1}{s} - \frac{\omega^2}{s^2} F(s)$$

for $s > 0$. Solving this for $F(s)$ gives $F(s) = \frac{s}{s^2 + \omega^2}$ for $s > 0$.

(7) Use induction. In the case $k = 0$,

$$F(s) = \int_0^\infty e^{-st} \, dt = \left[\frac{e^{-st}}{-s} \right]_{t=0}^{t=\infty} = \frac{1}{s}$$

for $s > 0$. Now suppose the formula holds for some $k \geq 0$. That is, assume we know that the Laplace transform of t^k is $k!/s^{k+1}$. Let $f(t) = t^{k+1}$. Then integrating by parts once gives

$$F(s) = \int_0^\infty t^{k+1} e^{-st} \, dt$$

$$= \left[\frac{e^{-st}}{-s} t^{k+1} \right]_{t=0}^{t=\infty} + \frac{k+1}{s} \int_0^\infty t^k e^{-st} \, dt$$

$$= \frac{k+1}{s} \frac{k!}{s^{k+1}}$$

$$= \frac{(k+1)!}{s^{k+2}},$$

so the formula holds for $n = k + 1$.

(9)

$$F(s) = \int_0^\infty \delta(t - b) e^{-st} \, dt = e^{-st} \Big|_{t=b} = e^{-bs}$$

12.5.2.

(ii)

$$\int_0^\infty \frac{df}{dt} e^{-st} \, dt = \left[f(t) e^{-st} \right]_{t=0}^{t=\infty} - \int_0^\infty f(t)(-s e^{-st}) \, dt$$

$$= s \int_0^\infty f(t) e^{-st} \, dt - f(0) = sF(s) - f(0)$$

187

(iv)

$$\int_0^\infty e^{bt} f(t) e^{-st}\, dt = \int_0^\infty f(t) e^{-(s-b)t}\, dt = F(s-b)$$

(vi)

$$\int_0^\infty t f(t) e^{-st}\, dt = -\int_0^\infty f(t) \frac{d}{ds}(e^{-st})\, dt = -\frac{d}{ds}\int_0^\infty f(t) e^{-st}\, dt = -\frac{dF}{ds}$$

(ix)

$$\int_0^\infty \int_0^t g(t-t') f(t')\, dt' e^{-st}\, dt = \int_0^\infty \int_0^t g(t-t') f(t') e^{-st}\, dt'\, dt$$

$$\text{(change order of integration)} \quad = \int_0^\infty \int_{t'}^\infty g(t-t') f(t') e^{-st}\, dt\, dt'$$

$$\text{(let } u = t - t') \quad = \int_0^\infty \int_0^\infty g(u) f(t') e^{-s(u+t')}\, du\, dt'$$

$$= \int_0^\infty f(t') e^{-st'} \int_0^\infty g(u) e^{-su}\, du\, dt'$$

$$= \int_0^\infty f(t') e^{-st'}\, dt' \int_0^\infty g(u) e^{-su}\, du$$

$$= F(s)G(s)$$

12.5.5. Let $F(x,s)$ denote the Laplace transform of u with respect to t. Then using property (iii) of the Laplace transform,

$$s^2 F(x,s) - su(x,0) - u_t(x,0) = c^2 F_{xx}(x,s),$$

so that the initial data gives

$$c^2 F_{xx}(x,s) = s^2 F(x,s) + (1-s)\sin(\pi x/l).$$

The boundary conditions imply that $F(0,s) = F(l,s) = 0$. For any s, the solution has the form $F(x) = A\sin(\pi x/l)$. Plugging this into the ODE, we find that

$$-c^2 \frac{\pi^2}{l^2} A = s^2 A + 1 - s.$$

Solving for A, we find

$$F(x,s) = \left(\frac{s-1}{s^2 + c^2\pi^2/l^2} \right) \sin(\pi x/l).$$

Using entries (12.5.3) and (12.5.4) in the table of Laplace transforms, one sees that the first factor in this expression is the Laplace transform of $\cos(\pi ct/l) - \frac{l}{\pi c}\sin(\pi ct/l)$. Thus

$$u(x,t) = \left(\cos(\pi ct/l) - \frac{l}{\pi c}\sin(\pi ct/l) \right) \sin(\pi x/l).$$

188

12.5.6 Let $F(x, s)$ denote the Laplace transform of u with respect to t. Then

$$s^2 F(x, s) - su(x, 0) - u_t(x, 0) = c^2 F_{xx}(x, s) + \frac{s}{s^2 + \omega^2} \sin \pi x,$$

and the initial data implies

$$s^2 F(x, s) = c^2 F_{xx}(x, s) + \frac{s}{s^2 + \omega^2} \sin \pi x.$$

The boundary conditions imply $F(0, s) = F(l, s) = 0$. The solution of this ODE in x is

$$F(x, s) = \frac{s}{(s^2 + \omega^2)(s^2 + c^2 \pi^2)} \sin(\pi x).$$

Since, for $\omega \neq c\pi$,

$$\frac{s}{(s^2 + \omega^2)(s^2 + c^2 \pi^2)} = \frac{s}{c^2 \pi^2 - \omega^2} \left(\frac{1}{s^2 + \omega^2} - \frac{1}{s^2 + c^2 \pi^2} \right)$$

is the Laplace transform of

$$\frac{1}{c^2 \pi^2 - \omega^2} \left(\cos \omega t - \cos c\pi t \right)$$

is follows that

$$u(x, t) = \frac{1}{c^2 \pi^2 - \omega^2} \left(\cos \omega t - \cos c\pi t \right) \sin(\pi x)$$

for $\omega \neq c\pi$. If $\omega = c\pi$, then

$$F(x, s) = \frac{s}{(s^2 + c^2 \pi^2)^2} \sin(\pi x) = -\frac{1}{2} \frac{d}{ds} \left(\frac{1}{s^2 + c^2 \pi^2} \right).$$

Using formula (12.5.4) and property (vi) it follows that

$$\frac{s}{(s^2 + c^2 \pi^2)^2} = -\frac{1}{2} \frac{d}{ds} \left(\frac{1}{s^2 + c^2 \pi^2} \right)$$

is the Laplace transform of $\frac{1}{2\pi c} t \sin(\pi c t)$ and therefore

$$u(x, t) = \frac{1}{2\pi c} t \sin(\pi c t) \sin(\pi x).$$

Chapter 13

Section 13.1

13.1.1. There is an error in the text; it should say $\frac{\partial \rho}{\partial t} = -\nabla \cdot \mathbf{J}$.

Answer: Differentiate (III) and then use (I) together with the identity $\nabla \cdot (\nabla \times \mathbf{F}) = 0$ to get

$$4\pi \frac{\partial \rho}{\partial t} = \frac{\partial}{\partial t} \nabla \cdot \mathbf{E} = \nabla \cdot \frac{\partial \mathbf{E}}{\partial t} = \nabla \cdot (c\nabla \times \mathbf{B} - 4\pi \mathbf{J}) = -4\pi \nabla \cdot \mathbf{J}.$$

13.1.3.

(a) Denote $\mathbf{K} = \mathbf{E} + \frac{1}{c}\frac{\partial \mathbf{A}}{\partial t}$. Then

$$\nabla \times \mathbf{K} = \nabla \times \mathbf{E} + \frac{1}{c}\frac{\partial}{\partial t}\nabla \times \mathbf{A} = \nabla \times \mathbf{E} + \frac{1}{c}\frac{\partial \mathbf{B}}{\partial t} = 0$$

by (I). Any vector field whose curl vanishes is the divergence of some scalar function (another well-known fact in vector analysis). Therefore there exists a scalar function u such that $\mathbf{K} = -\nabla u$.

(b) Substitute (a) into (I) and (III) and use the vector identity $\nabla \times (\nabla \times \mathbf{F}) = \nabla(\nabla \cdot \mathbf{F}) - \Delta \mathbf{F}$.

(c) λ is any scalar function. We can write $\mathbf{B} = \nabla \times \mathbf{A} = \nabla \times (\mathbf{A} + \nabla\lambda)$ because $\nabla \times \nabla\lambda = \mathbf{0}$. Furthermore,

$$\mathbf{E} = -\nabla u - \frac{1}{c}\frac{\partial \mathbf{A}}{\partial t} = -\nabla\left(u - \frac{1}{c}\frac{\partial \lambda}{\partial t}\right) - \frac{1}{c}\frac{\partial}{\partial t}(\mathbf{A} + \nabla\lambda).$$

(d) Given $\mathbf{A}^\bullet$ and $u^\bullet$, we want to find λ so that

$$\nabla \cdot (\mathbf{A}^\bullet + \nabla\lambda) + \frac{1}{c}\frac{\partial}{\partial t}\left(u^\bullet - \frac{1}{c}\frac{\partial \lambda}{\partial t}\right) = 0.$$

That is,

$$\frac{1}{c^2}\frac{\partial^2 \lambda}{\partial t^2} - \Delta\lambda = \nabla \cdot \mathbf{A}^\bullet + \frac{1}{c}\frac{\partial u^\bullet}{\partial t}.$$

We must solve this for λ. We can choose any initial condition $\not\equiv 0$ we please. The equation for λ is then solved using the inhomogeneous version of (9.2.3); see Theorem 3.4.1 for the one-dimensional case.

(e) Substitute the equation $\nabla \cdot \mathbf{A} + \frac{1}{c}\frac{\partial u}{\partial t} = 0$ into the result of part (b).

13.1.5. Consider for instance the last term in the equation above (13.1.8). Write $r = |\mathbf{x} - \mathbf{x}_0|$.

$$\frac{1}{t_0}\iint_S \mathbf{E}^0 dS = \frac{1}{t_0}\iint_{\{r = ct_0\}} \mathbf{E}^0(\mathbf{x})\, dS_{\mathbf{x}} = \int_0^{2\pi}\int_0^\pi \mathbf{E}^0(\mathbf{x})\, c^2 t_0 \sin\theta d\theta d\phi$$

where $\mathbf{x} = (x_0 + ct_0 \sin\theta\cos\phi,\ y_0 + ct_0\sin\theta\sin\phi,\ z_0 + ct_0\cos\theta)$. Differentiating this expression with respect to t_0, we get

$$\frac{\partial}{\partial t_0}\frac{1}{t_0}\iint_S \mathbf{E}^0 dS = c^2\int_0^{2\pi}\int_0^{\pi}\mathbf{E}^0(\mathbf{x})\sin\theta\, d\theta\, d\phi$$

$$+\, c^2 t_0\int_0^{2\pi}\int_0^{\pi}\left[c\sin\theta\cos\phi\frac{\partial\mathbf{E}^0}{\partial x} + c\sin\theta\sin\phi\frac{\partial\mathbf{E}^0}{\partial y} + c\cos\theta\frac{\partial\mathbf{E}^0}{\partial z}\right]\sin\theta\ d\theta\, d\phi.$$

The first integral comes from differentiating the t_0 directly, while the second integral comes from differentiating inside the argument of $\mathbf{E}_0$. Rewriting this formula in terms of surface integrals, we get

$$\frac{\partial}{\partial t_0}\frac{1}{t_0}\iint_S \mathbf{E}^0 dS = \frac{1}{t_0^2}\iint_S \mathbf{E}^0 dS + \frac{c}{t_0}\iint_S \frac{\partial\mathbf{E}^0}{\partial r}dS.$$

This implies (13.1.8). Formula (13.1.9) is left to the reader.

13.1.7. Starting from formula (13.1.8) for $\mathbf{E}(\mathbf{x}_0, t_0)$, consider $\lim_{t_0\searrow 0}\mathbf{E}(\mathbf{x}_0, t_0)$ as the sum of three terms. The first term in (13.1.8) is bounded by

$$\left|\frac{1}{4\pi ct_0}\iint_S \nabla\times\mathbf{B}^0\ dS\right| \le \frac{1}{4\pi ct_0}\left(\max|\nabla\times\mathbf{B}^0|\right)\left(4\pi c^2 t_0^2\right) \le (constant)t_0\ ,$$

which obviously tends to zero as $t_0\searrow 0$. The third term is similar. Thus the first and third terms both tend to zero. Therefore

$$\lim_{t_0\searrow 0}\mathbf{E}(\mathbf{x}_0, t_0) = 0 + \lim_{t_0\searrow 0}\frac{1}{4\pi c^2 t_0^2}\iint_S \mathbf{E}^0 dS\ + 0.$$

This is exactly the *average* of $\mathbf{E}^0$ on the sphere S and therefore it tends to $\mathbf{E}^0(\mathbf{x}_0)$ as the radius t_0 shrinks to zero. (Assume of course at least that $\mathbf{E}^0(\mathbf{x})$ is continuous at $\mathbf{x}_0$.) Therefore $\lim_{t_0\searrow 0}\mathbf{E}(\mathbf{x}_0, t_0) = \mathbf{E}^0(\mathbf{x}_0)$. The limit of $\mathbf{B}$ is left to the reader.

Section 13.2

13.2.1. Assume irrotational unforced flow; that is,

$$\frac{\partial\mathbf{v}}{\partial t} + (\mathbf{v}\cdot\nabla)\mathbf{v} = -\frac{1}{\rho}\nabla p, \quad \mathbf{v} = \nabla\phi, \quad p = f(\rho).$$

A vector identity says that $\nabla(|\mathbf{v}|^2) = 2(\mathbf{v}\cdot\nabla)\mathbf{v} + 2\mathbf{v}\times(\nabla\times\mathbf{v})$. But with $\mathbf{v} = \nabla\phi$, we have $\nabla\times\mathbf{v} = \nabla\times(\nabla\phi) = 0$ so that $(\mathbf{v}\cdot\nabla)\mathbf{v} = \nabla(\frac{1}{2}|\mathbf{v}|^2)$. Therefore the PDE becomes

$$\nabla\left\{\frac{\partial\phi}{\partial t} + \frac{1}{2}|\mathbf{v}|^2\right\} = -\frac{1}{\rho}\nabla p = -\frac{1}{\rho}f'(\rho)\nabla\rho.$$

Let g be defined by $g'(\rho) = f'(\rho)/\rho$. Then $\frac{1}{\rho}f'(\rho)\nabla\rho = g'(\rho)\nabla\rho = \nabla\{g(\rho)\}$. So integrating the PDE, we get

$$\frac{\partial\phi}{\partial t} + \frac{1}{2}|\mathbf{v}|^2 + g(\rho) = constant.$$

But

$$g(\rho) + constant = \int g'(\rho)d\rho = \int \frac{f'(\rho)}{\rho}d\rho = \int \frac{dp}{\rho}.$$

Section 13.3

13.3.2. We are assuming that $c_1 = c_2 = c = T = k = M = 1$, so that $u_{tt} = u_{xx}$ for $x \neq 0$. As in (13.3.2),

$$u(x,t) = \begin{cases} F(x-t) + G(x+t) & \text{for } x < 0 \\ H(x-t) + K(x+t) & \text{for } x > 0. \end{cases}$$

The given initial condition means that $F = f$ and $K = 0$. Thus

$$u(x,t) = \begin{cases} f(x-t) + G(x+t) & \text{for } x < 0 \\ H(x-t) & \text{for } x > 0, \end{cases}$$

where G is the reflected wave (going to the left) and H is the transmitted wave (going to the right). In this exercise we must find G and H. Put the formula for u into the jump conditions to get

$$H'(-t) - f'(-t) - G'(t) = f(-t) + G(t) + f''(-t) + G''(t) = H(-t) + H''(-t).$$

This is a double equality. Eliminate H as follows. Differentiating the last equality, we get

$$-H'(-t) - H'''(-t) = -f'(-t) + G'(t) - f'''(-t) + G'''(t).$$

Rewriting the first equality, we have

$$H'(-t) - f'(-t) - G'(t) = f(-t) + G(t) + f''(-t) + G''(t).$$

Differentiating twice, we get

$$H'''(-t) - f'''(-t) - G'''(t) = f''(-t) + G''(t) + f''''(-t) + G''''(t).$$

Combining the last three equations, it is easy to eliminate H entirely. After noticing the cancellation of the f' and f''' terms and reorganizing a bit, we obtain

$$G''''(t) + 2G'''(t) + 2G''(t) + 2G'(t) + G(t) = -f''''(-t) - 2f''(-t) - f(-t).$$

This is a fourth-order ODE to be solved for G. What are the four initial conditions for this ODE at $t = 0$? Well, $u(x,0) = f(x)$, $u_t(x,0) = -f(-x)$ so that $u(x,0+) = 0$, $u_t(x,0+) = 0$. Putting these into the previous expression for u in terms of f, G, H, we find that $G(0) = G'(0) = H(0) = H'(0) = 0$. It follows that $G(0) = G'(0) = G''(0) = G'''(0) = 0$. Now, to solve the fourth-order ODE for G, it is convenient (for brevity) to first denote $P(t) = -f(-t)$. Let D denote a derivative. Then the ODE is

$$(D^4 + 2D^3 + 2D^2 + 2D + 1)G = (D^4 + 2D^2 + 1)P,$$

which can be factored (just like a polynomial) as

$$(D^2 + 1)(D - 1)^2 G = (D^2 + 1)^2 P.$$

Notice the common factor. The initial conditions are all vanishing, as mentioned above. Thus we merely have to solve the second-order ODE

$$(D - 1)^2 G = (D^2 + 1)P, \quad G(0) = G'(0) = 0.$$

This last step is left to the reader.

13.3.4. Above equation (13.3.15) there are three terms. Call them I, II and III.

$$I = \left(l + \frac{1}{2}\right) \frac{i}{r} \left[e^{-ir} P_l(1) - e^{ir} P_l(-1)\right].$$

Now $P_l(1) = 1$ and $P_l(-1) = (-1)^l$. (See Section 10.6.) Therefore

$$I = \left(l + \frac{1}{2}\right) \frac{i}{r} \left[e^{-ir} - (-1)^l e^{ir}\right] = \left(l + \frac{1}{2}\right) \frac{i}{r} (-1)^l \sin\left(r - \frac{l\pi}{2}\right).$$

Now

$$|II| = \left|\left(\frac{i}{r}\right)^2 e^{-irs} P_l'(s)\Big|_{-1}^{1}\right| \leq \frac{1}{r^2}(|P_l'(1)| + |P_l'(-1)|) = \frac{const}{r^2}.$$

The third term is

$$|III| = \left|\left(\frac{i}{r}\right)^2 \int_{-1}^{1} e^{-irs} P_l''(s)ds\right| \leq \frac{1}{r^2} \int_{-1}^{1} |P_l''(s)|ds = \frac{const}{r^2}.$$

13.3.7. First solve the PDE in the whole of space without boundary conditions. Such a solution will only depend on $r = (x^2 + y^2 + (z - a)^2)^{1/2}$ and we are therefore led to an ODE. The ODE is easily solved to get $v = (1/r) \exp(\pm ikr)$. Finding the reflected wave means imposing the *outgoing* radiation condition at ∞. (See equation (13.3.10) in the text.) Hence $v = \frac{C}{r} e^{-ikr}$. It is straightforward to check by direct differentiation that this really solves the PDE $\Delta v = v_{rr} + (2/r)v_r = -k^2 v$ for $r \neq 0$. In fact, $C = -1/4\pi$ in order to get the delta function. [Notice that in case $k = 0$ it reduces exactly to the Green's function (12.2.1); see Exercise 7.2.1 for this case.] Thus

$$v = -\frac{1}{4\pi r} e^{-ikr}.$$

The next step is to use the method of reflection across the half-plane $z = 0$ as in Section 7.4 to get

$$v = -\frac{1}{4\pi} \left\{ \frac{1}{r} e^{ikr} - \frac{1}{r^*} e^{ikr^*} \right\},$$

where $r^* = (x^2 + y^2 + (z + a)^2)^{1/2}$ with a plus sign. This is accomplished exactly as in (7.4.1).

Section 13.4

13.4.2. The equation is

$$-\frac{d^2\psi}{dx^2} - Q\delta(x)\psi(x) = \lambda\psi(x).$$

If λ is an eigenvalue with an eigenfunction $\psi(x)$, then by definition $\int |\psi|^2 dx < \infty$. So we are looking for solutions $\neq 0$ that decay at infinity. For $x \neq 0$, the potential term is completely missing. So the solutions are exponentials if $\lambda < 0$, while the solutions do not decay if $\lambda \geq 0$ (because they are sines and cosines for $\lambda > 0$). So we can write $\lambda = -\beta^2 < 0$ and conclude that

$$\psi(x) = Ce^{-\beta|x|} \quad \text{for } x \neq 0.$$

Differentiating, we get

$$\psi'(x) = -C\beta e^{-\beta|x|}\text{sign}(x)$$

and

$$\psi''(x) = -C\beta e^{-\beta|x|}2\delta(x) + C\beta^2 e^{-\beta|x|}.$$

See (12.1.15) for an explanation of the last differentiation. Now insert these expressions into the ODE to get

$$0 = -\frac{d^2\psi}{dx^2} - Q\delta(x)\psi(x) - \beta^2\psi(x) = Ce^{-\beta|x|}\{-2\beta\delta(x) + \beta^2 + Q\delta(x) - \beta^2\}.$$

This is true if and only if $Q = 2\beta$. Thus the only eigenvalue and eigenfunction are

$$\lambda = -\frac{Q^2}{4} < 0, \quad \psi(x) = Ce^{-\sqrt{-\lambda}\,|x|}.$$

(The rest of the spectrum is continuous.)

Section 13.5

13.5.3.

(a) Proof that $\mathscr{E}$ is an invariant:

$$\frac{\partial \mathscr{E}}{\partial t} = \iiint \left(\mathbf{E}_1 \cdot \frac{\partial \mathbf{E}_1}{\partial t} + \cdots + \mathbf{B}_3 \cdot \frac{\partial \mathbf{B}_3}{\partial t}\right) d\mathbf{x}.$$

The first term in this integrand is

$$\mathbf{E}_1 \cdot \frac{\partial \mathbf{E}_1}{\partial t} = \mathbf{E}_1 \cdot (D_0\mathbf{E}_1 + \mathbf{A}_0 \times \mathbf{E}_1) = \mathbf{E}_1 \cdot D_0\mathbf{E}_1$$

$$= \mathbf{E}_1 \cdot \left(\frac{\partial \mathbf{B}_3}{\partial x_2} + \mathbf{A}_2 \times \mathbf{B}_3 - \frac{\partial \mathbf{B}_2}{\partial x_3} - \mathbf{A}_3 \times \mathbf{B}_2\right).$$

194

Write out the other 5 terms in the same way. You will find that all the 12 triple products vanish because of the identity $\mathbf{A} \cdot (\mathbf{B} \times \mathbf{C}) = (\mathbf{A} \times \mathbf{B}) \cdot \mathbf{C}$. You will end up with the integrand

$$\mathbf{E}_1 \cdot \left(\frac{\partial \mathbf{B}_3}{\partial x_2} - \frac{\partial \mathbf{B}_2}{\partial x_3}\right) + \mathbf{E}_2 \cdot \left(\frac{\partial \mathbf{B}_1}{\partial x_3} - \frac{\partial \mathbf{B}_3}{\partial x_1}\right) + \mathbf{E}_3 \cdot \left(\frac{\partial \mathbf{B}_2}{\partial x_1} - \frac{\partial \mathbf{B}_1}{\partial x_2}\right)$$

$$+\mathbf{B}_1 \cdot \left(\frac{\partial \mathbf{E}_2}{\partial x_3} - \frac{\partial \mathbf{E}_3}{\partial x_2}\right) + \mathbf{B}_2 \cdot \left(\frac{\partial \mathbf{E}_3}{\partial x_1} - \frac{\partial \mathbf{E}_1}{\partial x_3}\right) + \mathbf{B}_3 \cdot \left(\frac{\partial \mathbf{E}_1}{\partial x_2} - \frac{\partial \mathbf{E}_2}{\partial x_1}\right)$$

$$= \frac{\partial}{\partial x_1}(\mathbf{B}_2 \cdot \mathbf{E}_3 - \mathbf{B}_3 \cdot \mathbf{E}_2) + \frac{\partial}{\partial x_2}(\mathbf{B}_3 \cdot \mathbf{E}_1 - \mathbf{B}_1 \cdot \mathbf{E}_3) + \frac{\partial}{\partial x_3}(\mathbf{B}_1 \cdot \mathbf{E}_2 - \mathbf{B}_2 \cdot \mathbf{E}_1).$$

(b) For the momentum, take for instance its first component.

$$\frac{\partial \mathscr{P}_1}{\partial t} = \frac{\partial}{\partial t} \iiint (\mathbf{B}_2 \cdot \mathbf{E}_3 - \mathbf{B}_3 \cdot \mathbf{E}_2)d\mathbf{x}$$

and calculate in a similar way.

13.5.4. Let's start with the first equation, which is $D_0\mathbf{B}_1 = D_3\mathbf{E}_2 - D_2\mathbf{E}_3$. Using the definitions of the D operators, this is written as

$$\frac{\partial \mathbf{B}_1}{\partial t} - \mathbf{A}_0\mathbf{B}_1 + \mathbf{B}_1\mathbf{A}_0 = \frac{\partial \mathbf{E}_2}{\partial x_3} + \mathbf{A}_3\mathbf{E}_2 - \mathbf{E}_2\mathbf{A}_3 - \left(\frac{\partial \mathbf{E}_3}{\partial x_2} + \mathbf{A}_2\mathbf{E}_3 - \mathbf{E}_3\mathbf{A}_2\right).$$

Multiply this equation on the left by G^{-1} and on the right by G, where $G = G(x,t)$ is a unitary 2×2 matrix function with determinant $=1$. By definition, $\mathbf{B}'_k = G^{-1}\mathbf{B}_kG$ and $\mathbf{E}'_k = G^{-1}\mathbf{E}_kG$. Take for instance the left side of the resulting equation, which is

$$G^{-1}D_0\mathbf{B}_1G = G^{-1}\frac{\partial \mathbf{B}_1}{\partial t}G - G^{-1}\mathbf{A}_0\mathbf{B}_1G + G^{-1}\mathbf{B}_1\mathbf{A}_0G.$$

The last term on the right is

$$G^{-1}\mathbf{B}_1\mathbf{A}_0G = G^{-1}\mathbf{B}_1GG^{-1}\mathbf{A}_0G = \mathbf{B}'_1\left[\mathbf{A}'_0 + G^{-1}\frac{\partial G}{\partial t}\right]$$

by definition of $\mathbf{B}'_1$ and $\mathbf{A}'_0$. Similarly, the next-to-last term is

$$-G^{-1}\mathbf{A}_0\mathbf{B}_1G = -G^{-1}\mathbf{A}_0GG^{-1}\mathbf{B}_1G = -\left[\mathbf{A}'_0 + G^{-1}\frac{\partial G}{\partial t}\right]\mathbf{B}'_1.$$

The first term on the right is, by the differentiation rules for matrices,

$$G^{-1}\frac{\partial \mathbf{B}_1}{\partial t}G = \frac{\partial(G^{-1}\mathbf{B}_1G)}{\partial t} - G^{-1}\mathbf{B}_1\frac{\partial G}{\partial t} + G^{-1}\frac{\partial G}{\partial t}G^{-1}\mathbf{B}_1G$$

$$= \frac{\partial \mathbf{B}'_1}{\partial t} - \mathbf{B}'_1G^{-1}\frac{\partial G}{\partial t} + G^{-1}\frac{\partial G}{\partial t}\mathbf{B}'_1.$$

When we combine these three expressions, the terms involving G cancel and we end up with

$$\frac{\partial \mathbf{B}_1}{\partial t} - \mathbf{A}_0 \mathbf{B}_1 + \mathbf{B}_1 \mathbf{A}_0 = \frac{\partial \mathbf{B}'_1}{\partial t} - \mathbf{A}'_0 \mathbf{B}'_1 + \mathbf{B}'_1 \mathbf{A}'_0 \ .$$

Thus the term $D_0 \mathbf{B}_1$ looks identical in the new variables after the gauge transformation. All of the other terms in the Yang-Mills equations transform in the same manner.

13.5.9. Start with the telegraph equation $u_{tt} - c^2 u_{xx} + u_t = 0$ and make the transformation $u = e^{-t/2} v$. Then $u_t = e^{-t/2}[v_t - \frac{1}{2} v]$ and $u_{tt} = e^{-t/2}[v_{tt} - v_t + \frac{1}{4} v]$ by direct calculation. Therefore the PDE becomes, after dividing by $e^{-t/2}$,

$$v_{tt} - c^2 v_{xx} - \tfrac{1}{4} v = 0.$$

This is the Klein-Gordon equation with $m = i/2$. Using (13.5.14), the source function for the telegraph equation is

$$S(x, t) = \frac{1}{2c} e^{-t/2} J_0\left(\frac{i}{2}\sqrt{t^2 - x^2/c^2}\right) \quad \text{for } |x| \le ct$$

for $t \ge 0$, and $S(x, t) = 0$ for $|x| \ge ct$.

13.5.11. Let

$$\mathbf{A} = i\begin{pmatrix} \alpha_1 & \alpha_2 + i\alpha_3 \\ \alpha_2 - i\alpha_3 & -\alpha_1 \end{pmatrix}, \quad \mathbf{B} = i\begin{pmatrix} \beta_1 & \beta_2 + i\beta_3 \\ \beta_2 - i\beta_3 & -\beta_1 \end{pmatrix}.$$

Then calculate, by straightforward matrix multiplication, $\mathbf{AB} - \mathbf{BA} =$

$$2i\begin{pmatrix} (\alpha_2 \beta_3 - \alpha_3 \beta_2) & (\alpha_3 \beta_1 - \alpha_1 \beta_3) + i(\alpha_1 \beta_2 - \alpha_2 \beta_1) \\ (\alpha_3 \beta_1 - \alpha_1 \beta_3) - i(\alpha_1 \beta_2 - \alpha_2 \beta_1) & -(\alpha_2 \beta_3 - \alpha_3 \beta_2) \end{pmatrix}.$$

Therefore the vectors corresponding to these matrices are

$$\mathbf{A} \sim \begin{pmatrix} \alpha_1 \\ \alpha_2 \\ \alpha_3 \end{pmatrix}, \quad \mathbf{B} \sim \begin{pmatrix} \beta_1 \\ \beta_2 \\ \beta_3 \end{pmatrix}, \quad \mathbf{A} \times \mathbf{B} \sim \begin{pmatrix} \alpha_2 \beta_3 - \alpha_3 \beta_2 \\ \alpha_3 \beta_1 - \alpha_1 \beta_3 \\ \alpha_1 \beta_2 - \alpha_2 \beta_1 \end{pmatrix}.$$

Chapter 14

Section 14.1

14.1.2 From the PDE $(1+t)u_t + xu_x = 0$, the characteristic ODE is

$$\frac{dx}{dt} = \frac{x}{1+t}.$$

Thus $dx/x = dt/(1+t)$, so that $\log x = \log(1+t)+(\text{constant})$, and $x = C(1+t)$ where C is a constant. Along any characteristic curve (that satisfies the ODE), u is a constant. Therefore

$$u(C(1+t),t) = u(C,0) = \phi(C)$$

where the initial condition is $u(x,0) = \phi(x)$. So

$$u(x,t) = \phi\left(\frac{x}{1+t}\right).$$

If $\phi(x) = x^5$, then

$$u(x,t) = \left(\frac{x}{1+t}\right)^5.$$

14.1.3. Since $\phi(x) = x$, equation (14.1.10) for the characteristic lines becomes $x - x_0 = tx_0$. So $x_0 = x/(1+t)$ and (14.1.11) becomes $u(x,t) = x_0$ or

$$u(x,t) = \frac{x}{1+t}.$$

The characteristic lines are given by the equation above with "slope" $dx/dt = x_0$ and passing through $(x_0, 0)$. Thus the picture is a "fan" of lines all of which intersect at the point $(0, -1)$. See Figure 26.

14.1.6. Start with (14.1.12), which reads

$$u(x,t) = \frac{1}{2t^2} + \frac{x}{t} - \frac{\sqrt{1+4tx}}{2t^2}.$$

Differentiation yields the following expressions.

$$u_t = -\frac{1}{t^3} - \frac{x}{t^2} - \frac{x}{t^2\sqrt{1+4tx}} + \frac{1+4tx}{t^3\sqrt{1+4tx}}.$$

$$u_x = \frac{1}{t} - \frac{1}{t\sqrt{1+4tx}}.$$

Then the product uu_x is

$$uu_x = \left\{\frac{1}{2t^2} + \frac{x}{t} - \frac{\sqrt{1+4tx}}{2t^2}\right\}\left\{\frac{1}{t} - \frac{1}{t\sqrt{1+4tx}}\right\}$$

$$= \frac{1}{t^3} - \frac{1}{2t^3\sqrt{1+4tx}} + \frac{x}{t^2} - \frac{x}{t^2\sqrt{1+4tx}} - \frac{1+4tx}{2t^3\sqrt{1+4tx}}.$$

197

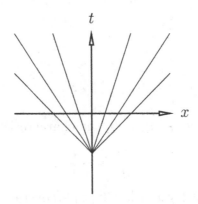

Figure 26: Characteristics for the problem in Exercise 14.1.3.

Hence

$$u_t + uu_x = \frac{1}{t^3\sqrt{1+4tx}}\{-xt+1+4tx-1-xt-2tx\} = 0.$$

This whole calculation is valid only in the region $1 + 4tx \geq 0$, which is between the two branches of the hyperbola.

14.1.9. Differentiation of $u(x,t) = \phi(z)$ leads to $u_t = \phi'(z)z_t$ and $a(u)u_x = a(\phi(z))\phi'(z)z_x$ whence

$$u_t + a(u)u_x = \phi'(z)[z_t + a(\phi(z))z_x].$$

On the other hand, $x - z = ta(\phi(z))$ leads to $-z_t = a(\phi(z)) + ta'(\phi(z))\phi'(z)z_t$ and $1 - z_x = ta'(\phi(z))\phi'(z)z_x$. Thus

$$z_t = \frac{-a(\phi(z))}{1+ta'(\phi(z))\phi'(z)}, \quad z_x = \frac{1}{1+ta'(\phi(z))\phi'(z)}.$$

Substitution of these expressions for z_t and z_x leads to

$$u_t + a(u)u_x = \phi'(z)\left\{\frac{-a(\phi(z)) + a(\phi(z))\cdot 1}{1+ta'(\phi(z))\phi'(z)}\right\} = 0.$$

14.1.10. We are supposed to find the solution of the PDE $u_t + uu_x = 0$ with the initial condition

$$u(x,0) = \phi(x) = \begin{cases} 1 & \text{for } x \leq 0 \\ 1-x & \text{for } 0 \leq x \leq 1 \\ 0 & \text{for } 1 \leq x. \end{cases}$$

Use formulas (14.1.10) and (14.1.11). The characteristic lines are given by (14.1.10), namely, $x - x_0 = t\phi(x_0)$. Thus the characteristic line that passes through the point $(x_0, 0)$ should be

$$x - x_0 = \begin{cases} t & \text{if } x_0 \leq 0 \\ t(1-x_0) & \text{if } 0 \leq x_0 \leq 1 \\ 0 & \text{if } 1 \leq x_0. \end{cases}$$

198

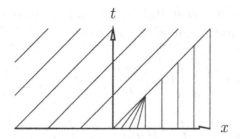

Figure 27: Characteristics for the problem in Exercise 14.1.10.

Clearly the characteristic lines collide at the point $(1,1)$, as is particularly evident in Figure 27.

However *until they collide*, for $0 \leq t < 1$ we can solve for x_0 and for $u(x,t)$ to get

$$
x_0 = \begin{cases} x - t \\ \dfrac{x-t}{1-t} \ , \\ x \end{cases} \quad u(x,t) = \phi(x_0) = \begin{cases} 1 & \text{for } x \leq t \\ 1 - \dfrac{x-t}{1-t} = \dfrac{1-x}{1-t} & \text{for } t \leq x \leq 1 \\ 0 & \text{for } 1 \leq x. \end{cases}
$$

Now at $t = 1$ the solution is

$$
u(x,1) = \begin{cases} 1 & \text{for } x < 1 \\ 0 & \text{for } x > 1. \end{cases}
$$

Therefore for $t > 1$ it must look just like Figure 14.1.7. The "slope" of the shock wave has to be given by the Rankine-Hugoniot formula (with $a(u) = u$, $A(u) = \frac{1}{2}u^2$):

$$
\frac{dx}{dt} = s = \frac{\frac{1}{2}(u^+)^2 - \frac{1}{2}(u^-)^2}{u^+ - u^-} = \frac{1}{2}(u^+ + u^-) = \frac{1}{2}(0 + 1) = \frac{1}{2}.
$$

(See the discussion of Example 14.1.5 in the text.) The entropy condition (14.1.18) on the shock wave becomes $1 = u^- > s > u^+ = 0$, which is true since $s = \frac{1}{2}$.

14.1.11 We begin with

$$
\int_0^\infty \int_{-\infty}^\infty [u\psi_t + A(u)\psi_x]\, dx\,dt = 0.
$$

Choosing $\psi(x,t) = \phi(x)\,\rho(t)$, we have

$$
\int_0^\infty \rho'(t) \int_{-\infty}^\infty u\,\phi(x)\, dx\,dt + \int_0^\infty \rho(t) \int_{-\infty}^\infty A(u)\,\phi'(x)\, dx\,dt = 0.
$$

Now choose $\phi(x)$ to be

$$
\phi(x) = \begin{cases} 0 & \text{for } x < a - \epsilon \\ 1 & \text{for } a < x < b \\ 0 & \text{for } b + \epsilon < x \end{cases}
$$

199

and smooth in the intervals $[a - \epsilon, a]$ and $[b, b + \epsilon]$. For simplicity we take $\phi(x)$ to be linear with slope $1/\epsilon$ in $[a - \epsilon, a]$ and linear with slope $-1/\epsilon$ in $[b, b + \epsilon]$. Then

$$\int_0^\infty \rho'(t) \left\{ \int_a^b u \; dx + \int_{a-\epsilon}^a \left(1 + \frac{x-a}{\epsilon}\right) u \; dx - \int_b^{b+\epsilon} \left(1 - \frac{x-b}{\epsilon}\right) u \; dx \right\} dt$$

$$+ \int_0^\infty \rho(t) \left\{ \frac{1}{\epsilon} \int_{a-\epsilon}^a A(u) dx - \frac{1}{\epsilon} \int_b^{b+\epsilon} A(u) dx \right\} dt = 0.$$

Letting $\epsilon \to 0$, we get

$$\int_0^\infty \rho'(t) \left[\int_a^b u \; dx \right] dt + \int_0^\infty \rho(t) \left[A(u(a,t)) - A(u(b,t)) \right] dt = 0.$$

Since $\rho(t)$ is arbitrary, we have

$$-\frac{d}{dt} \left[\int_a^b u \; dx \right] + A(u(a,t)) - A(u(b,t)) = 0$$

for all a, b, t.

Section 14.2

14.2.2. Solution 1. Let $I = \int_{-\infty}^{+\infty} (xu - 3tu^2) dx$. Then

$$\frac{dI}{dt} = \int (xu_t - 6tuu_t - 3u^2) dx = \int \left\{ (x - 6tu)(-u_{xxx} - 6uu_x) - 3u^2 \right\} dx,$$

where the KdV equation (14.2.1) has been inserted for u_t. Now

$$-(x - 6tu)u_{xxx} = -[(x - 6tu)u_{xx}]_x + (1 - 6tu_x)u_{xx} = -[(x - 6tu)u_{xx}]_x + [u_x - 3tu_x^2]_x.$$

Furthermore,

$$-(x - 6tu)(6uu_x) = -(x - 6tu)(3u^2)_x = -[(x - 6tu)3u^2]_x + (1 - 6tu_x)3u^2$$
$$= -[(x - 6tu)3u^2]_x + 3u^2 - [6tu^3]_x.$$

Hence

$$\frac{dI}{dt} = \int_{-\infty}^{+\infty} J_x \; dx = J \Big|_{-\infty}^{+\infty},$$

where

$$J = -(x - 6tu)(u_{xx} + 3u^2) - 6tu^3 + u_x - 3tu_x^2.$$

It is assumed that all the terms in J vanish as $|x| \to \infty$. Hence $dI/dt \equiv 0$ so that I is an invariant.

Solution 2. We already know that $M = \int u^2 dx$ is an invariant. That is, M is a constant. So $I = \int xu \, dx - 3Mt$. So

$$\frac{dI}{dt} = \int xu_t \, dx - 3M = -\int x(u_{xxx} + 6uu_x)dx - 3M = -\int x(u_{xx} + 3u^2)_x dx - 3M$$

$$= \int_{-\infty}^{\infty} \left\{ -[x(u_{xx} + 3u^2)]_x + (u_{xx} + 3u^2) \right\} dx - 3M = 3M - 3M = 0.$$

14.2.4.

(a) $P(f) = -2f^3 + cf^2 + 2af + b$. The equation for traveling waves is $(f')^2 = P(f)$. So it is required that $P(f) \geq 0$. This ODE is integrated by writing

$$\frac{df}{dx} = f'(x) = \pm\sqrt{P(f(x))}.$$

Thus $df/\sqrt{P(f)} = \pm dx$ so that

$$\int_{f(0)}^{f(x)} \frac{dr}{\sqrt{P(r)}} = \pm x.$$

This is the implicit formula for $f(x)$.

(b) Suppose f_1 is a simple zero of P. That is, $P(f_1) = 0$, $P'(f_1) \neq 0$. Suppose now that $f(x_1) = f_1$. Since $f'(x) = \pm\sqrt{P(f(x))}$, we have $f'(x_1) = \pm\sqrt{P(f_1)} = 0$. Differentiating, we get

$$f''(x) = \pm\tfrac{1}{2}[P(f(x))]^{-\frac{1}{2}} P'(f(x)) f'(x) = \tfrac{1}{2}P'(f(x)).$$

Thus $f''(x_1) = \tfrac{1}{2}P'(f_1) \neq 0$. Because $f'(x_1) = 0$ and $f''(x_1) \neq 0$, the function $f(x)$ has a local maximum or minimum at x_1.

(c) P is a cubic polynomial. Hence one can always find a, b so that it has three real zeros. For example, one can choose $b = 0$ and $a > 0$. Clearly $P(f) \to \pm\infty$ as $f \to \mp\infty$. Call the zeros $f_1 < f_2 < f_3$. So $P > 0$ in the intervals $(-\infty, f_1)$ and (f_2, f_3).

(d) Returning to the second order solitary wave equation $-cf + f'' + 3f^2 = a$, we may rewrite it as a system

$$f' = g$$
$$g' = cf - 3f^2 + a$$

The above calculations imply that $g^2 = P(f)$. This describes a curve in the (f, g)-plane which, when the polynomial P has three distinct roots, consists of one unbounded curve together with one closed curve. See Figure 29. Any initial data (f, g) along the closed curve leads to a periodic solution, that oscillates between $f = f_2$ and $f = f_3$.

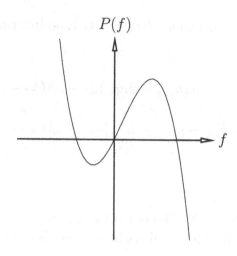

Figure 28: Graph of $P(f)$.

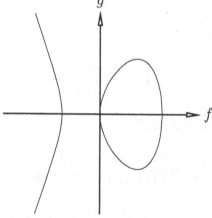

Figure 29: Graph of $g^2 = P(f)$.

(e) Write $P(r) = -2(r - f_1)(r - f_2)(r - f_3)$ where $f_1 < f_2 < f_3$. Notice that, within the integral, $f_2 \leq r \leq f_3$ so that $0 \leq f_3 - r \leq f_3 - f_2$. Thus we can find θ so that $f_3 - r = (f_3 - f_2)\sin^2 \theta$ with $0 \leq \theta \leq \pi/2$. So we substitute

$$r = f_3 + (f_2 - f_3)\sin^2 \theta, \quad dr = 2(f_2 - f_3)\sin\theta\cos\theta\, d\theta.$$

Now

$$\frac{1}{2}P(r) = -(r - f_1)(r - f_2)(r - f_3)$$

$$= -\{(f_3 - f_1) + (f_2 - f_3)\sin^2\theta\}\{(f_3 - f_2) + (f_2 - f_3)\sin^2\theta\}\{(f_2 - f_3)\sin^2\theta\}$$

$$= -(f_3 - f_1)\left\{1 - \frac{f_3 - f_2}{f_3 - f_1}\sin^2\theta\right\}(f_3 - f_2)\{1 - \sin^2\theta\}(f_3 - f_2)\{-\sin^2\theta\}$$

$$= (f_3 - f_1)(f_3 - f_2)^2\{1 - k^2\sin^2\theta\}\cos^2\theta\sin^2\theta$$

where $k^2 = (f_3 - f_2)/(f_3 - f_1)$.

202

Upon taking the square root and dividing, we get a lot of cancellation;

$$\frac{dr}{\sqrt{P(r)}} = \frac{-\sqrt{2}\,d\theta}{\sqrt{f_3 - f_1}\sqrt{1 - k^2\sin^2\theta}}.$$

Therefore,

$$x = \int_{f_3}^{f(x)} \frac{dr}{\sqrt{P(r)}} = \frac{-\sqrt{2}}{\sqrt{f_3 - f_1}} \int_0^{\phi(x)} \frac{d\theta}{\sqrt{1 - k^2\sin^2\theta}},$$

taking the $+$ sign for simplicity.

Now the elliptic integral of the first kind is defined as

$$E(\phi) = \int_0^{\phi} \frac{d\theta}{\sqrt{1 - k^2\sin^2\theta}}.$$

So $-\frac{1}{\sqrt{2}} x\sqrt{f_3 - f_1} = E(\phi(x))$ and

$$f(x) = f_3 + (f_2 - f_3)\sin^2\phi(x) = f_2 + (f_3 - f_2)\cos^2\phi(x).$$

One defines the *cnoidal function* cn by the equation $\operatorname{cn}(E(\phi)) = \cos\phi$. Then $f(x) = f_2 + (f_3 - f_2)\operatorname{cn}^2(E(\phi))$ or

$$f(x) = f_2 + (f_3 - f_2)\operatorname{cn}^2\left(-\frac{\sqrt{f_3 - f_1}}{2}x\right).$$

14.2.5. Write (14.2.4) as $\psi_{xx} + (\lambda + u)\psi = 0$ with $\lambda = \lambda(t)$ and $\psi = \psi(x,t)$. Differentiating, we have

$$\psi_{xxx} + (\lambda + u)\psi_x + u_x\psi = 0, \quad \psi_{xxt} + (\lambda + u)\psi_t + (\lambda_t + u_t)\psi = 0.$$

Now h is defined by

$$h = \psi_t + 2(u - 2\lambda)\psi_x - u_x\psi$$

so that

$$h_{xx} = \psi_{xxt} + 2(u - 2\lambda)\psi_{xxx} + 3u_x\psi_{xx} - u_{xxx}\psi.$$

Thus

$$\begin{aligned}
(\psi h_x - \psi_x h)_x &= \psi h_{xx} - \psi_{xx} h \\
&= \psi\{\psi_{xxt} + 2(u - 2\lambda)\psi_{xxx} + 3u_x\psi_{xx} - u_{xxx}\psi\} \\
&\quad - \psi_{xx}\{\psi_t + 2(u - 2\lambda)\psi_x - u_x\psi\}.
\end{aligned}$$

Substituting the previous expressions for ψ_{xxt} and ψ_{xxx}, many terms cancel and we find

$$\begin{aligned}
(\psi h_x - \psi_x h)_x &= -\psi^2\{\lambda_t + u_t + (\lambda + u)u_x + 2(u - 2\lambda)u_x + 3(\lambda + u)u_x + u_{xxx}\} \\
&= -\psi^2\{\lambda_t + u_t + 6uu_x + u_{xxx}\} = -\lambda_t\psi^2
\end{aligned}$$

using the KdV equation (14.2.1) at the last step. This proves (14.2.6).

14.2.8.

(a) Let us first consider the transmission coefficient $T(t, k)$. Since h/ψ is a function of t only,

$$\frac{h}{\psi} = \lim_{x \to -\infty} \frac{\psi_t - 2(-u + 2\lambda)\psi_x - u_x\psi}{\psi} = \frac{T_t e^{-ikx} - 2(0 + 2\lambda)(-ikTe^{-ikx}) - 0}{Te^{-ikx}}$$

so that

$$\frac{h}{\psi} = \frac{T_t + 4\lambda ikT}{T} = \frac{T_t}{T} + 4\lambda ik.$$

But from the text,

$$\frac{h}{\psi} = \frac{R_t - 4\lambda ikR}{R} = 4\lambda ik.$$

Thus $T_t = 0$ and T is a constant.

(b) Now consider the normalizing constants. Let $\lambda = -\kappa_n^2 < 0$ be an eigenvalue and let $\psi_n(x)$ be the corresponding eigenfunction, where $\int_{-\infty}^{\infty} \psi_n^2 \, dx = 1$. By (14.2.6),

$$-(\kappa_n^2)_t \int_{-\infty}^{\infty} \psi_n^2 \, dx \; + \; (\psi_n h_{nx} - \psi_{nx} h_n)\Big|_{-\infty}^{\infty} = 0.$$

Since ψ_n and h_n decay exponentially at infinity, we deduce that κ_n is a constant. Now (14.2.6) implies that $\psi_n h_{nx} - \psi_{nx} h_n$ is independent of x, and therefore so is $h_n/\psi_n \equiv \alpha_n(t)$. For simplicity of notation, we drop the subscripts n. Thus

$$\alpha\psi = h = \psi_t - 2(2\lambda - u)\psi_x - u_x\psi.$$

Multiplying by ψ, we get

$$\alpha\psi^2 - (\tfrac{1}{2}\psi^2)_t + (u\psi^2)_x = 4(-\lambda + u)\psi\psi_x$$

$$= 4\left(-2\lambda - \frac{\psi_{xx}}{\psi}\right)\psi\psi_x = (-4\lambda\psi^2 - 2\psi_x^2)_x.$$

Thus

$$\alpha\psi^2 = (\tfrac{1}{2}\psi^2)_t - (u\psi^2 + 4\lambda\psi^2 + 2\psi_x^2)_x.$$

Since u and ψ vanish at infinity, integration yields

$$\alpha = \alpha \int_{-\infty}^{\infty} \psi_n^2 \, dx = \tfrac{1}{2}\frac{d}{dt}\int_{-\infty}^{\infty} \psi_n^2 \, dx = 0.$$

So also

$$0 = h = h_n(x, t) = \psi_{nt} + 2(2\kappa_n^2 + u)\psi_{nx} - u_x\psi_n.$$

As $x \to +\infty$, $u \to 0$ and $\psi_n \sim c_n(t)e^{-\kappa_n x}$, so that in the limit

$$\frac{dc_n}{dt}e^{-\kappa_n x} + 2(2\kappa_n^2 + 0)c_n(-\kappa_n e^{-\kappa_n x}) - 0 = 0$$

or

$$\frac{dc_n}{dt} - 4\kappa_n^3 c_n = 0.$$

14.2.10. The Fourier transform is

$$U(k,t) = \int_{-\infty}^{\infty} u(x,t)e^{-ikx}dx$$

[see (12.3.4)]. From the PDE $u_t + u_{xxx} = 0$, property (i) of Fourier transforms yields $U_t + (ik)^3 U = 0$. Thus $U(k,t) = U(k,0)\exp(ik^3 t)$. Therefore, substituting and switching the order of integration (which is the "inversion principle" of the Fourier transform), we get

$$u(x,t) = \int_{-\infty}^{\infty} U(k,0)e^{ik^3 t}e^{ikx}\frac{dk}{2\pi}$$

$$= \int_{-\infty}^{\infty} \left(\int_{-\infty}^{\infty} u(s,0)e^{-iks}ds \right) e^{ik^3 t + ikx}\frac{dk}{2\pi}$$

$$= \int_{-\infty}^{\infty} \phi(s) \left\{ \int_{-\infty}^{\infty} e^{ik(x-s)+ik^3 t}\frac{dk}{2\pi} \right\} ds = \int_{-\infty}^{\infty} \phi(s)\, D(x-s)\, ds,$$

where

$$D(\xi,t) = \int_{-\infty}^{\infty} e^{ik\xi + ik^3 t}\frac{dk}{2\pi}.$$

Changing variables $l = (3t)^{1/3}k$ yields

$$D(\xi,t) = \int_{-\infty}^{\infty} \exp\left[il\xi(3t)^{-1/3} + i\frac{1}{3}l^3 \right] \left(\frac{3}{t}\right)^{1/3}\frac{dl}{2\pi} = \left(\frac{1}{3t}\right)^{1/3} A\left(\left(\frac{1}{3t}\right)^{1/3}\xi \right).$$

Using the notion of convolution from Section 12.3,

$$u(\cdot,t) = D(\cdot,t) * \phi(\cdot).$$

That is, at each time t, the solution u is the convolution of D and ϕ.

14.2.12. Following the method of Section 2.4, look first for a solution Q of the PDE $Q_t + Q_{xxx} = 0$ with $Q(x,0) = 1$ for $x > 0$ and $Q(x,0) = 0$ for $x < 0$. This initial condition is the Heaviside function (Section 12.1). Exercise 14.2.11 motivates looking for Q in the form

$$Q(x,t) = g(p) \quad \text{where} \quad p = \frac{x}{t^{\frac{1}{3}}}.$$

Differentiation gives

$$Q_t = \frac{-x}{3t^{4/3}}g'(p), \quad Q_x = \frac{1}{t^{1/3}}g'(p), \quad Q_{xxx} = \frac{1}{t}g'''(p).$$

Thus

$$0 = Q_t + Q_{xxx} = \frac{-x}{3t^{4/3}}g'(p) + \frac{1}{t}g'''(p) = \frac{1}{t}\left\{ -\frac{p}{3}g'(p) + g'''(p) \right\}.$$

So we have the ODE $g''' = (p/3)g'$. Given any solution, we have $Q(x,t) = g(x/t^{1/3})$. By ODE theory, this equation has solutions with three arbitrary constants of integration. It can be shown that these constants are determined by the initial condition for Q (omitted).

Now $S \equiv Q_x$ satisfies the same PDE with the initial condition $S(x,0) = \delta(x)$, the delta "function" (see Section 12.1). Thus it follows that the solution with initial condition $\phi(x)$ is $u(x,t) = \int_{-\infty}^{\infty} S(x-y,t)\phi(y)dy$. Furthermore, the function S is given by

$$S(x,t) = \frac{\partial}{\partial x} g\left(\frac{x}{t^{\frac{1}{3}}}\right) = \frac{1}{t^{\frac{1}{3}}} g'\left(\frac{x}{t^{\frac{1}{3}}}\right).$$

Section 14.3

14.3.2. Assuming y is a single-valued function of x, we will look for $y = u(x)$ that minimizes the length. That is, we minimize $E[u] = \int_0^1 \sqrt{1 + (du/dx)^2}\,dx$ subject to the constraints $\int_0^1 u\,dx = A$, $u(0) = a$, $u(1) = b$. Let $F(p) = \sqrt{1+p^2}$ so that $E[u] = \int_0^1 F(u_x)dx$. Furthermore, let $A[u] = \int_0^1 u\,dx$. By the method of Lagrange multipliers, there is a constant λ such that

$$0 = \frac{d}{d\epsilon}\{E[u + \epsilon v] - \lambda A[u + \epsilon v]\} = 0 \quad \text{for all } v.$$

Thus

$$\frac{d}{dx}F(u_x) = \frac{d}{dx}\frac{u_x}{\sqrt{1+u_x^2}} = \lambda \cdot 1 \ .$$

This is the Euler equation, which is to be solved subject to the three constraints. Integrating once yields $F(u_x) = \lambda x + c_1$ for some constant c_1. Taking the inverse function, $u_x = F^{-1}(\lambda x + c_1)$. The inverse function is easy to find, namely,

$$\text{if} \quad q = F(p) = \frac{p}{\sqrt{1+p^2}}, \quad \text{then} \quad p = F^{-1}(q) = \frac{q}{\sqrt{1-q^2}}.$$

Thus

$$\frac{du}{dx} = u_x = \frac{\lambda x + c_1}{\sqrt{1 - (\lambda x + c_1)^2}}.$$

Integration yields

$$u = c_2 - \frac{1}{\lambda}\sqrt{1 - (\lambda x + c_1)^2}$$

where c_2 is another constant. With the notation $y = u(x)$, this can be written as $\lambda^2(y-c_2)^2 = 1 - (\lambda x + c_1)^2$ or $\lambda^2(x^2 + y^2) = $ linear terms. This is the equation of a circle! Thus we conclude that the shortest curve is the arc of the unique circle that passes through $(0,a)$ and $(1,b)$ and has underlying area A. See Figure 30. These three constraints uniquely determine the three constants c_1, c_2, λ.

14.3.5.

(a) Just use the rules of differentiation as follows. Equation (14.3.8) is

$$\left(u_x(1 + u_x^2 + u_y^2)^{-1/2}\right)_x + \left(u_y(1 + u_x^2 + u_y^2)^{-1/2}\right)_y = 0.$$

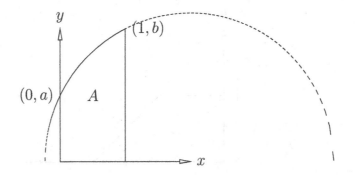

Figure 30: The shortest curve between two given points with given area below is an arc of a circle.

For brevity, write $p = 1 + u_x^2 + u_y^2$. Then, carrying out the derivatives, we get

$$0 = u_x \left(-\tfrac{1}{2}p^{-3/2}\right)(2u_x u_{xx} + 2u_y u_{yx}) + u_{xx} \, p^{-1/2}$$
$$+ u_y \left(-\tfrac{1}{2}p^{-3/2}\right)(2u_x u_{xy} + 2u_y u_{yy}) + u_{yy} \, p^{-1/2}.$$

Multiply by $p^{3/2}$ to get

$$0 = -u_x(u_x u_{xx} + u_y u_{yx}) + u_{xx}(1 + u_x^2 + u_y^2) - u_y(u_x u_{xy} + u_y u_{yy}) + u_{yy}(1 + u_x^2 + u_y^2)$$
$$= -2u_x u_y u_{xy} + u_y^2 u_{xx} + u_x^2 u_{yy} + u_{xx} + u_{yy}.$$

Thus

$$(1 + u_y^2)u_{xx} - 2u_x u_y u_{xy} + (1 + u_x^2)u_{yy} = 0.$$

14.3.7. Let

$$I[y] = \int_0^2 (y')^2 (1 + y')^2 \, dx.$$

The problem is to minimize $I[y]$ subject to the two boundary constraints $y(0) = 1$ and $y(2) = 0$. Using the notation of equation (14.3.4), $F(y') = (y')^2(1 + y')^2 = (y' + y'^2)^2$. So

$$\frac{\partial F}{\partial y'} = 2y' + 6y'^2 + 4y'^3.$$

So equation (14.3.6) becomes $0 = [2y' + 6y'^2 + 4y'^3]'$. Thus $y' + 3y'^2 + 2y'^3 = $ constant, so that $y' = $ constant. This is satisfied by *any* straight line $y = ax + b$. The two boundary constraints imply that $y = 1 - \frac{x}{2}$. For this line, $I[y] = \frac{1}{8}$.

However, we can make I smaller. In fact, we can make $I = 0$ as follows. (See Figure 31.) Let

$$y = \begin{cases} 1 & \text{for } 0 \le x \le 1 \\ 2 - x & \text{for } 1 \le x \le 2. \end{cases}$$

207

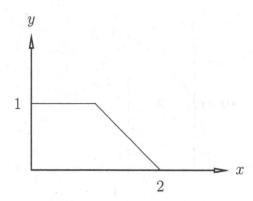

Figure 31: A minimizer for the problem in Exercise 14.3.7.

Then $y' = 0$ or -1, so that $I[y] = 0$. Clearly 0 is the minimum value of $I[y]$ since $I[y]$ is always non-negative. To find many more minimizers, let n be a positive integer and let

$$u_n(x) = \begin{cases} 1 - \dfrac{i}{n} & \text{for } \dfrac{2i}{n} \le x \le \dfrac{2i+1}{n} \\ 1 - \dfrac{i}{n} - \left(x - \dfrac{2i+1}{n}\right) & \text{for } \dfrac{2i+1}{n} \le x \le \dfrac{2i+2}{n} \end{cases}$$

for $i = 0, 1, \ldots, n - 1$. (The sketch for the case $n = 4$ is shown in Figure 32.) Then the

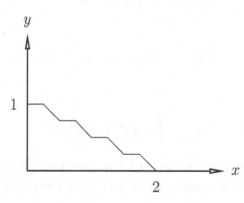

Figure 32: Another minimizer for the problem in Exercise 14.3.7.

derivative $u'_n(x) = 0$ or -1 in each little subinterval, so that $I[u_n] = 0$.

14.3.8. Let

$$H[f] = \iiint f(\mathbf{v}) \log f(\mathbf{v}) d\mathbf{v}, \quad E[f] = \iiint \frac{1}{2} |\mathbf{v}|^2 f(\mathbf{v}) d\mathbf{v},$$

$$\mathbf{M}[f] = \iiint \mathbf{v} f(\mathbf{v}) d\mathbf{v}, \quad m[f] = \iiint f(\mathbf{v}) d\mathbf{v}.$$

The integrations are taken over the whole of three-dimensional space. Now for any test function g,

$$\frac{d}{d\epsilon} H[f + \epsilon g]\Big|_{\epsilon=0} = \frac{d}{d\epsilon} \iiint (f + \epsilon g) \log(f + \epsilon g) d\mathbf{v}$$

$$= \iiint (f \log f)' \, g d\mathbf{v} = \iiint (1 + \log f) g d\mathbf{v}.$$

Similarly, for the other three functionals we get

$$\iiint \frac{1}{2}|\mathbf{v}|^2 g d\mathbf{v}, \quad \iiint \mathbf{v} g d\mathbf{v}, \quad \iiint g d\mathbf{v}.$$

Therefore, using Lagrange multipliers, any minimum must satisfy

$$\iiint \left\{ \log f(\mathbf{v}) + \lambda_1 \tfrac{1}{2}|\mathbf{v}|^2 + \boldsymbol{\lambda}_2 \cdot \mathbf{v} + \lambda_3 + 1 \right\} g(\mathbf{v}) d\mathbf{v} = 0$$

for all test functions $g(\mathbf{v})$. Hence $\log f(\mathbf{v}) + \lambda_1 \tfrac{1}{2}|\mathbf{v}|^2 + \boldsymbol{\lambda}_2 \cdot \mathbf{v} + \lambda_3 + 1 = 0$ so that

$$f(\mathbf{v}) = \exp\left\{ -\lambda_1 \tfrac{1}{2}|\mathbf{v}|^2 - \boldsymbol{\lambda}_2 \cdot \mathbf{v} - \lambda_3 - 1 \right\}$$

for some constants $\lambda_1, \boldsymbol{\lambda}_2, \lambda_3$.

We require $\lambda_1 > 0$ in order that $E[f] < \infty$. We complete the square in the exponent as

$$-\frac{1}{2}\lambda_1 |\mathbf{v}|^2 - \boldsymbol{\lambda}_2 \cdot \mathbf{v} - \lambda_3 - 1 = -\frac{1}{2}\lambda_1 \left| \mathbf{v} + \frac{\boldsymbol{\lambda}_2}{\lambda_1} \right|^2 + d,$$

where d is a constant.

Now for the minimum state, the momentum vanishes, so that (for $j = 1, 2, 3$)

$$0 = M_j[f] = \iiint v_j f(\mathbf{v}) d\mathbf{v} = \iiint v_j \exp\left\{ -\frac{1}{2}\lambda_1 \left| \mathbf{v} + \frac{\boldsymbol{\lambda}_2}{\lambda_1} \right|^2 + d \right\} d\mathbf{v}$$

$$= e^d \iiint \left(w_j + \frac{(\boldsymbol{\lambda}_2)_j}{\lambda_1} \right) \exp\left\{ -\frac{1}{2}\lambda_1 |\mathbf{w}|^2 \right\} d\mathbf{w}$$

where $\mathbf{w} = \mathbf{v} + \boldsymbol{\lambda}_2/\lambda_1$. Now

$$\int_{-\infty}^{\infty} w_j e^{-\lambda_1 w_j^2/2} dw_j = 0$$

because the integrand is an odd function of w_j. Therefore

$$0 = \frac{(\boldsymbol{\lambda}_2)_j}{\lambda_1} \iiint e^{-\lambda_1 |\mathbf{w}|^2/2} d\mathbf{w}.$$

Thus $(\boldsymbol{\lambda}_2)_j = 0$ for $j = 1, 2, 3$. Therefore $\boldsymbol{\lambda}_2 = \mathbf{0}$. Next, the mass is

$$1 = m[f] = e^{-\lambda_3 - 1} \iiint e^{-\frac{1}{2}\lambda_1 |\mathbf{v}|^2} d\mathbf{v} = \left(\frac{2\pi}{\lambda_1} \right)^{\frac{3}{2}} e^{-\lambda_3 - 1}.$$

Therefore (see Exercise 2.4.6)

$$f(\mathbf{v}) = \left(\frac{\lambda_1}{2\pi}\right)^{\frac{3}{2}} e^{-\frac{1}{2}\lambda_1 |\mathbf{v}|^2}.$$

Now the energy is

$$E[f] = \left(\frac{\lambda_1}{2\pi}\right)^{\frac{3}{2}} \frac{1}{2} \iiint |\mathbf{v}|^2 e^{-\frac{1}{2}\lambda_1 |\mathbf{v}|^2} \, d\mathbf{v} = \frac{1}{\pi^{\frac{3}{2}}\lambda_1} \iiint |\mathbf{w}|^2 e^{-|\mathbf{w}|^2} \, d\mathbf{w},$$

where $\mathbf{w} = \sqrt{\lambda_1/2}\, \mathbf{v}$. The last integral can be explicitly calculated using the identity

$$|\mathbf{w}|^2 e^{-|\mathbf{w}|^2} = -\frac{1}{2} \sum_{j=1}^{3} w_j \frac{\partial}{\partial w_j} \left(e^{-|\mathbf{w}|^2}\right) = \frac{\partial}{\partial w_j} \left(-\frac{1}{2} w_j e^{-|\mathbf{w}|^2}\right) + \frac{3}{2} e^{-|\mathbf{w}|^2}.$$

Thus

$$\iiint |\mathbf{w}|^2 e^{-|\mathbf{w}|^2} \, d\mathbf{w} = \frac{3}{2} \iiint e^{-|\mathbf{w}|^2} \, d\mathbf{w} = \frac{3}{2} \pi^{\frac{3}{2}}.$$

So

$$E = E[f] = \frac{1}{\pi^{\frac{3}{2}}\lambda_1} \frac{3}{2} \pi^{\frac{3}{2}} = \frac{3}{2\lambda_1}$$

and $\lambda_1 = 3/(2E)$. We conclude that

$$f(\mathbf{v}) = \left(\frac{3}{4\pi E}\right)^{\frac{3}{2}} \exp\left\{-\frac{3}{4E} |\mathbf{v}|^2\right\}.$$

14.3.10.

(a) The action is

$$A[u] = \iint \left(\frac{1}{2} u_x u_t + u_x^3 - \frac{1}{2} u_{xx}^2\right) dx\, dt.$$

Consider the variation

$$A[u + \epsilon v] = \iint \left(\frac{1}{2}(u_x + \epsilon v_x)(u_t + \epsilon v_t) + (u_x + \epsilon v_x)^3 - \frac{1}{2}(u_{xx} + \epsilon v_{xx})^2\right) dx\, dt.$$

Taking the derivative with respect to ϵ and then setting $\epsilon = 0$ yields

$$0 = \iint \left\{\tfrac{1}{2}(u_t v_x + u_x v_t) + 3u_x^2 v_x - u_{xx} v_{xx}\right\} dx\, dt.$$

This is supposed to be valid for all "test" functions v. Integrating by parts and assuming as usual some vanishing conditions at infinity, we get

$$0 = \iint \left\{-\tfrac{1}{2}u_{tx} - \tfrac{1}{2}u_{xt} - 3(u_x^2)_x - u_{xxxx}\right\} v \, dx\, dt.$$

Hence

$$u_{tx} + 6u_x u_{xx} + u_{xxxx} = 0.$$

Section 14.4

14.4.1. In the equation $u_{xx} + \lambda \sin u = 0$, both $u(x)$ and λ are regarded as unknowns. Multiplying by u_x leads to $\frac{1}{2}u_x^2 - \lambda \cos u = \text{constant}$. Since $u(x) + \pi$ is another solution for $-\lambda$, we may as well take $\lambda = \mu^2 > 0$. Letting $\alpha = u(0)$ and recalling that $u_x(0) = 0$, we have

$$\tfrac{1}{2}u_x^2 = \lambda(\cos u - \cos \alpha).$$

Thus $\cos u \geq \cos \alpha$. Assuming for simplicity that $0 \leq \alpha \leq \pi$, it follows that $|u(x)| \leq \alpha$. The equation can be integrated in terms of $\int du/\sqrt{\cos u - \cos \alpha}$ but it is more convenient to change variables as follows.

Let $k = \sin \frac{\alpha}{2}$ and introduce a new variable ϕ by

$$k \sin \phi = \sin \tfrac{u}{2} .$$

Then $k\phi_x \cos \phi = \frac{1}{2}u_x \cos \frac{u}{2}$. Squaring and using the expression for u_x^2 yields

$$k^2 \phi_x^2 \cos^2 \phi = \frac{\lambda}{2}(\cos u - \cos \alpha) \cos^2 \frac{u}{2}.$$

But $\cos u = 1 - 2\sin^2 \frac{u}{2} = 1 - 2k^2 \sin^2 \phi$ and $\cos \alpha = 1 - 2\sin^2 \frac{\alpha}{2} = 1 - 2k^2$, so that $\cos u - \cos \alpha = 2k^2 \cos^2 \phi$. Thus

$$k^2 \phi_x^2 \cos^2 \phi = \lambda k^2 \cos^2 \phi \cos^2 \frac{u}{2}$$

and

$$\phi_x^2 = \lambda \cos^2 \frac{u}{2} = \lambda \left(1 - \sin^2 \frac{u}{2}\right) = \lambda(1 - k^2 \sin^2 \phi).$$

Hence

$$\frac{d\phi}{\sqrt{1 - k^2 \sin^2 \phi}} = \pm\mu \; dx, \qquad \mu = \lambda^2.$$

For simplicity, consider only the $+$ sign. Integration yields

$$\int_{\phi(0)}^{\phi(x)} \frac{dr}{\sqrt{1 - k^2 \sin^2 r}} = \mu x.$$

This is an elliptic integral of the first kind. The standard elliptic integral is

$$E(\phi; k) = \int_0^\phi \frac{dr}{\sqrt{1 - k^2 \sin^2 r}}.$$

Now $\sin \phi(0) = \frac{1}{k}\sin \frac{u(0)}{2} = 1$, so that $\phi(0)$ equals $\pi/2$ plus a multiple of 2π. At the other end $x = l$, we have $0 = u_x^2(l) = 2\lambda(\cos u(l) - \cos \alpha)$, so that $\cos u(l) = \cos \alpha = \cos u(0)$, $\sin^2 \phi(l) = \sin^2 \phi(0) = 1$, and $\phi(l)$ equals $\pi/2$ plus a multiple of π. Thus

$$\phi(0) = \frac{\pi}{2} + 2\pi p, \quad \phi(l) = \frac{\pi}{2} + \pi q$$

211

where p and q are integers.

Now we can plot μ as a function of k. Indeed, taking $x = l$ yields

$$\mu = \frac{1}{l} \int_{\frac{\pi}{2}+2\pi p}^{\frac{\pi}{2}+\pi q} \frac{dr}{\sqrt{1-k^2\sin^2 r}} = \frac{1}{l} \int_0^{n\pi} \frac{dr}{\sqrt{1-k^2\sin^2 r}} = \frac{n}{l} \int_0^{\pi} \frac{dr}{\sqrt{1-k^2\sin^2 r}}$$

where $n = q - 2p$ is an integer. Thus

$$\mu = \frac{n}{l} E(\pi;k) \qquad (n = 1, 2, 3, \dots).$$

Let's consider this equation as implicitly defining $k = \sin\frac{u(0)}{2}$ as a function of $\mu = \sqrt{\lambda}$. Because $E(\pi, 0) = \pi$, the point $(\frac{n}{l}\pi, 0)$ is a solution in the μk-plane for $n = 1, 2, \dots$. Each such point generates a bifurcation curve. Now

$$\frac{d\mu}{dk} = \frac{n}{l} \frac{\partial E(\pi, k)}{\partial k} > 0$$

for $k > 0$. Therefore each bifurcation curve bends to the right. Furthermore, for small k, we can use a Taylor expansion to approximately evaluate

$$E(\pi;k) = \int_0^{\pi} \left(1 + \frac{k^2}{2}\sin^2 r + \dots\right) dr = \pi + \frac{k^2}{2}\frac{\pi}{2} + \dots$$

so that

$$\mu = \frac{n\pi}{l} \left(1 + \frac{k^2}{4} + \dots\right).$$

Thus each curve looks like a parabola near $k = 0$; this is the "pitchfork" bifurcation. Another observation is that as $k \nearrow 1$, $E(\pi;k) \nearrow \infty$ and $\mu \nearrow \infty$. Thus the bifurcation curves in the μk-plane are asymptotic to the lines $k = \pm 1$. In terms of the original parameters λ and $\alpha = u(0)$, this means that the curves look like parabolas near $\alpha = 0$ and are asymptotic to $\mu = \pm\pi$. See Figure 33.

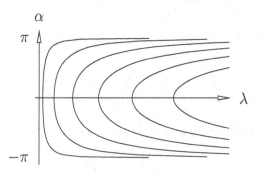

Figure 33: Bifurcation diagram for Example 14.4.1.

14.4.3. Start with equation (14.4.5):

$$l = \frac{1}{\sqrt{2}} \int_0^{\alpha(p)} \frac{du}{\sqrt{F(\alpha(p)) - F(u)}} \equiv \beta(p)$$

212

for $0 < p < A$. Here $F' = f$ and $F(0) = 0$ with f as in Figure 14.4.8. From Figure 14.4.9, $\alpha(A) = b$ so that

$$\beta(A) = \frac{1}{\sqrt{2}} \int_0^b \frac{du}{\sqrt{F(b) - F(u)}}.$$

But a Taylor expansion near $u = b$ shows that $F(u) - F(b) = f(b)(u-b) + O(u-b)^2$, whence $\sqrt{F(u) - F(b)} = O(|u - b|)$ as $u \to b$. Therefore the preceding integral diverges because of the behavior of the integrand near b. So $\beta(p) \to \infty$ as $p \to A$.

Moreover, a curve in Figure 14.4.9 connects the origin $(0,0)$ with the point $(a, 0)$, so that $\alpha(0) = a$ and $F(\alpha(0)) = \frac{1}{2}0^2 + F(\alpha(0)) = E = \frac{1}{2}0^2 + F(0) = 0$. Thus

$$\beta(0) = \frac{1}{\sqrt{2}} \int_0^{\alpha(0)} \frac{du}{\sqrt{-F(u)}}.$$

But a Taylor expansion near $u = 0$ shows that $F(u) = O(u^2)$. That is, $\sqrt{|F(u)|} \leq c|u|$ for small $|u|$. So the integral diverges because of the behavior of the integrand near 0. So $\beta(p) \to \infty$ as $p \to 0$. Finally, $\beta(p)$, being a continuous function, must have a finite minimum β_0 somewhere between 0 and A.

Proof that the minimum is unique: Since the points $(0, p)$ and $(\alpha(p), 0)$ lie on the same curve, we have $F(\alpha(p)) = H(\alpha(p), 0) = H(0, p) = \frac{1}{2}p^2 + F(0)$ so that $f(\alpha(p))\alpha'(p) = p$. Since $a < \alpha(p) < b$, we have $f(\alpha(p)) > 0$ and $\alpha'(p)) > 0$. Therefore, in order to prove that the minimum of $\beta(p)$ is unique, all we need to show is that the function

$$S(\alpha) = \int_0^\alpha \frac{du}{\sqrt{F(\alpha) - F(u)}}$$

has only one point where $S'(\alpha) = 0$. To do that, it is enough show that $S''(\alpha) > 0$ at any point where $S'(\alpha) = 0$.

It is tricky to differentiate S directly, so we first change variables in the integral, writing

$$u = \alpha \sin \psi, \quad du = \alpha \cos \psi \, d\psi.$$

Thus

$$S(\alpha) = \int_0^{\pi/2} \frac{\alpha \cos \psi \, d\psi}{\sqrt{F(\alpha) - F(\alpha \sin \psi)}}.$$

Writing for brevity $G(\alpha) = G(\alpha, \psi) = F(\alpha) - F(\alpha \sin \psi)$, and letting $'$ denote a derivative with respect to α, we have

$$S'(\alpha) = \int_0^{\pi/2} \left[G^{-\frac{1}{2}} - \frac{\alpha}{2} G^{-\frac{3}{2}} G' \right] \cos \psi \, d\psi,$$

$$S''(\alpha) = \int_0^{\pi/2} \left[-G^{-\frac{3}{2}} G' + \frac{3\alpha}{4} G^{-\frac{5}{2}} (G')^2 - \frac{\alpha}{2} G^{-\frac{3}{2}} G'' \right] \cos \psi \, d\psi.$$

The latter integral may be rewritten as

$$\int_0^{\pi/2} \left\{ \frac{1}{2} G^{-\frac{3}{2}} (G' - \alpha G'') + \frac{3}{4\alpha} G^{-\frac{5}{2}} (\alpha G' - 2G)^2 - \frac{3}{\alpha} \left(-\frac{\alpha}{2} G^{-\frac{3}{2}} G' + G^{-\frac{1}{2}} \right) \right\} \cos \psi \, d\psi.$$

213

The second term is obviously non-negative while the third term coincides with $-4S'(\alpha)$. Therefore

$$S''(\alpha) + \frac{3}{\alpha}S'(\alpha) \geq \int_0^{\pi/2} \frac{1}{2}G^{-\frac{3}{2}}(G' - \alpha G'')\cos\psi\, d\psi.$$

Returning to the variable u, we have

$$S''(\alpha) + \frac{3}{\alpha}S'(\alpha) \geq \int_0^{\pi/2} \frac{f(\alpha) - f(\alpha\sin\psi)\sin\psi - \alpha f'(\alpha) - \alpha f'(\alpha\sin\psi)\sin^2\psi}{2G^{3/2}}\cos\psi\, d\psi$$

$$= \frac{1}{2\alpha^2}\int_0^\alpha [F(\alpha) - F(u)]^{-\frac{3}{2}}\, [\alpha f(\alpha) - u f(u) - \alpha^2 f'(\alpha) + u^2 f(u)]\, du.$$

It is convenient to let $h(u) = 2F(u) - uf(u)$. Then

$$S''(\alpha) + \frac{3}{\alpha}S'(\alpha) \geq \frac{1}{2\alpha^2}\int_0^\alpha [F(\alpha) - F(u)]^{-\frac{3}{2}}\, [\alpha h'(\alpha) - u h'(u)]\, du \qquad \text{(S-12)}$$

and

$$S'(\alpha) = \frac{1}{2}\int_0^\alpha [F(\alpha) - F(u)]^{-\frac{3}{2}}[h(\alpha) - h(u)]\, du.$$

Since we are assuming $f(u) = -u(u-a)(u-b) = -u^3 + (a+b)u^2 - abu$, we easily calculate that $h(u) = u^3[\frac{1}{2}u - \frac{1}{3}(a+b)]$ decreases from $u = 0$ to its minimum at $u = (a+b)/2$. Thus if $\alpha < (a+b)/2$, then $h(\alpha) - h(u) < 0$ in the integrand for $S'(u)$, so that $S'(\alpha) < 0$.

Furthermore, $uh'(u) = 2u^3[u - (a+b)/2]$ is negative for $0 < u < (a+b)/2$ and positive for $u > (a+b)/2$. Now let α be a point where $S'(\alpha) = 0$. Then $\alpha \geq (a+b)/2$ and $\alpha h'(\alpha) > uh'(u)$ for all $0 < u < \alpha$. By (S-12) it follows that $S''(\alpha) > 0$. As discussed above, this conclusion implies that the minimum of $\beta(p)$ is unique.

Section 14.5

14.5.2. Let $q = \frac{1}{2}u^2 + \frac{1}{2}w^2 + p + gz$. The second and third equations in (14.5.3) are

$$u_t + uu_x + wu_z + p_x = 0, \qquad w_t + uw_x + ww_z + p_z + g = 0.$$

Adding ww_x and uu_z to these equations, we get

$$u_t + q_x = w(w_x - u_z) = w\omega, \qquad w_t + q_z = -u(w_x - u_z) = -u\omega.$$

Therefore

$$\left(\frac{1}{2}u^2 + \frac{1}{2}w^2 + gz\right)_t = uu_t + ww_t = -uq_x - wq_z = -(uq)_x - (wq)_z.$$

Integrating over z from 0 to $\eta(x,t)$, we get

$$\left[\int_0^\eta \left(\frac{1}{2}u^2 + \frac{1}{2}w^2 + gz\right)dz\right]_t - \eta_t \left(\frac{1}{2}u^2 + \frac{1}{2}w^2 + gz\right)\Big|_0^\eta = -\left[\int_0^\eta uq\, dz\right]_x + \eta_x uq\Big|_0^\eta - wq\Big|_0^\eta.$$

214

The last term evaluated on the bottom $z = 0$ vanishes because $w = 0$. On the top $S = \{z = \eta(x,t)\}$, we have the boundary condition $w = \eta_t + u\eta_x$, so that almost all the terms on the top cancel. The other boundary condition on S is that $p = P_0$ is a constant. Therefore

$$\frac{d}{dt}\int_0^L \int_0^\eta \left(\frac{1}{2}u^2 + \frac{1}{2}w^2 + gz\right) dz\,dx = -\int_0^L \eta_t P_0\,dx = -P_0\frac{dm}{dt},$$

where $m = \int_0^L \int_0^\eta dz\,dx = \int_0^L \eta\,dx$. Now

$$\frac{dm}{dt} = \int_0^L \eta_t\,dx = \int_0^L (w - \eta_x u)\,dx = \int_0^L (-\psi_x - \eta_x \psi_y)\,dx = \psi(x,\eta(x))\Big|_0^L = 0.$$

Thus $\mathscr{E}$ is a constant, and so is m.

14.5.3. Let

$$Q = \frac{1}{2}\Psi_X^2 + \frac{1}{2}\Psi_z^2 + gz + p + \Gamma(\Psi) = \frac{1}{2}(u-c)^2 + \frac{1}{2}w^2 + gz + p + \Gamma(\Psi),$$

where $\Gamma' = \gamma$. Then

$$Q_x = (u-c)u_x + ww_x - [(u-c)u_x + wu_z] + \gamma(\Psi)\Psi_x = w\omega - \gamma(\Psi)w = 0$$

and

$$Q_z = (u-c)u_z + ww_z + g - [(u-c)w_x + ww_z + g] + \gamma(\Psi)\Psi_z = -(u-c)\omega + \gamma(\Psi)(u-c) = 0.$$

So Q is a constant throughout D_η. In particular, Q is a constant on S. Now on S we know that $p = P_0$ and $\Psi = 0$. So two of the terms in Q are constants anyway. Thus $\frac{1}{2}\Psi_X^2 + \frac{1}{2}\Psi_z^2 + gz$ is a constant on S.

14.5.6. For $\epsilon = 0$, equations (14.5.14) and (14.5.15) reduce to

$$u_x + w_z = 0 \tag{S-13}$$
$$u_t = -p_x \tag{S-14}$$
$$0 = p_z \tag{S-15}$$

in $\{0 < z < 1\}$, together with the boundary conditions

$$w = 0 \text{ on } \{z = 0\} \tag{S-16}$$
$$p = \eta \text{ on } \{z = 1\} \tag{S-17}$$
$$w = \eta_t \text{ on } \{z = 1\}. \tag{S-18}$$

By (S-15) and (S-17), $p = \eta(x,t)$ is independent of z. By (S-14), $u_t = -\eta_x$ is also independent of z. So by (S-13), $w_{zt} = -u_{xt} = -u_{tx} = \eta_{xx}(x,t)$. So by (S-18), $\eta_{tt}(x,t) = w_t(x,1,t) = \eta_{xx}(x,t)$. This is the wave equation for η.

Changes in Numbering from the First Edition

The numbering in this solution manual corresponds to the Second Edition of the textbook. Below is a list of changes in numbering from the First Edition. The notation 5-7 → 6-8 indicates that Exercises 5,6 and 7 in the First Edition are now Exercises 6,7 and 8 in the Second Edition.

Section	Change in Numbering
1.2	5-7 → 6-8, 8-11 → 10-13
1.4	4-6 → 5-7
2.5	2 → 4
4.2	3 → 4
7.4	26 → 25
11.6	4-10 → 2-8
12.1	4-10 → 6-12
13.1	3(b)-(f) → 3(a)-(e)
14.1	1-9 → 3-11